Microsoft HoloLens By Example

Create immersive augmented reality experiences

Joshua Newnham

BIRMINGHAM - MUMBAI

Microsoft HoloLens By Example

First published: August 2017

Production reference: 1290817

Published by Packt Publishing Ltd.
Livery Place
35 Livery Street
Birmingham
B3 2PB, UK.
ISBN 978-1-78712-626-8

www.packtpub.com

Credits

Author
Joshua Newnham

Reviewers
Vangos Pterneas
Jason M. Odom

Commissioning Editor
Amarabha Banerjee

Acquisition Editor
Reshma Raman

Content Development Editor
Aditi Gour

Technical Editor
Lydia Vazhayil

Copy Editor
Shaila Kusanale

Project Coordinator
Ritika Manoj

Proofreader
Safis Editing

Indexer
Tejal Daruwale Soni

Graphics
Jason Monteiro

Production Coordinator
Shantanu Zagade

About the Author

Joshua Newnham is a lead design technologist at Method, an experience design studio, with a focus on the intersection of design and artificial intelligence, working on a variety of projects ranging from wearables to assist medical staff during a surgery to a health assistant chatbot. Prior to this, Joshua was a technical director at Masters of Pie, a digital production studio focused on designing and developing virtual reality (VR) and augmented reality (AR) experiences, and premonitory developing tools for engineers and creatives.

Other roles before this included head of mobile for the digital agency--Razorfish--and cofounder of one of London's first mobile production studios--ubinow--acquired by Havas Workclub, where he lead many projects for major global brands such as Universal Pictures, Diageo, and adidas.

Joshua's interest lies in designing and building digital solutions that make our lives a little easier and more enjoyable, exploring new ways in which we can interact with computers, and designing and developing systems that are more intuitive, intelligent and social, and thus better integrated into our daily lives.

I would like to first and foremost thank my loving and patient wife and son for their continued support, patience, and encouragement throughout the long process of writing this book. Thanks also to the Masters of Pie and Method teams for their generosity with their equipment--obviously a critical component for this book. Of course, thanks to Aditi Gour and the Packt team for their enthusiasm in the content, flexibility, and assistance in helping push this book over the finish line. Finally, I would like to pay tribute to the team at Microsoft responsible for the HoloLens platform (and the likes) for not settling for a world where our interaction with computers is dominated by a screen, but wanting it immersed into our world.

About the Reviewer

Vangos Pterneas helps innovative companies increase their revenue using motion technology and virtual reality. He is an expert in Kinect, HoloLens, Oculus Rift, and HTC Vive.

Microsoft has awarded him with the title of Most Valuable Professional for his technical contributions to the open source community. Vangos runs LightBuzz Inc, collaborating with clients from all over the world. He's also the author of *Getting Started with HTML5 WebSocket Programming* and *The Dark Art of Freelancing*.

www.PacktPub.com

For support files and downloads related to your book, please visit www.PacktPub.com. Did you know that Packt offers eBook versions of every book published, with PDF and ePub files available? You can upgrade to the eBook version at www.PacktPub.com and as a print book customer, you are entitled to a discount on the eBook copy. Get in touch with us at service@packtpub.com for more details.

At www.PacktPub.com, you can also read a collection of free technical articles, sign up for a range of free newsletters and receive exclusive discounts and offers on Packt books and eBooks.

https://www.packtpub.com/mapt

Get the most in-demand software skills with Mapt. Mapt gives you full access to all Packt books and video courses, as well as industry-leading tools to help you plan your personal development and advance your career.

Why subscribe?

- Fully searchable across every book published by Packt
- Copy and paste, print, and bookmark content
- On demand and accessible via a web browser

Customer Feedback

Thanks for purchasing this Packt book. At Packt, quality is at the heart of our editorial process. To help us improve, please leave us an honest review on this book's Amazon page at `https://www.amazon.com/dp/1787126269`

If you'd like to join our team of regular reviewers, you can e-mail us at `customerreviews@packtpub.com`. We award our regular reviewers with free eBooks and videos in exchange for their valuable feedback. Help us be relentless in improving our products!

Table of Contents

Preface

This book is about developing mixed reality applications for the HoloLens platform, specifically focusing on core topics unique to mixed reality applications and diving deep into areas such as hologram placement, modes of interaction, and sharing experiences across devices; these are delivered through a series of examples, each focusing on illustrating one or more of these core topics. This focus on core topics differentiates this book from others, along with illustrating one of these through non-trivial and, hopefully, inspiring examples.

After completing this book, you will have a good grasp of the HoloLens platform and, more importantly, be inspired and motivated to continue your journey of exploring and experimenting with envisioning new ways in which we can interact with our computers with platforms such as Microsoft HoloLens.

What this book covers

Chapter 1, *Enhancing Reality*, introduces readers to the concepts of mixed reality, contrasting it against augmented reality (AR) and virtual reality (VR), including a discussion about the opportunities it offers in creating more natural and immersive experiences.

Chapter 2, *Tagging the World Using DirectX*, begins walking through setting up the environment, and then walks the user through building an augmented reality application that leverages the Microsoft Cognitive Services Computer Vision API to tag recognized content within the users' field of view, specifically, recognizing faces.

Chapter 3, *Assistant Item Finder Using DirectX*, that explains how to build on top of the previous example by allowing the user to pin tagged items and have the system assist the user navigate back to them.

Chapter 4, *Building Paper Trash Ball in Unity*, walks the reader through the development of a popular smartphone game--Paper Toss--but unlike the traditional game, the setting will be the user's physical environment.

Chapter 5, *Building Paper Trash Ball Using HoloToolkit in Unity*, is a continuation of the last chapter, but looks at reimplementing the project using HoloToolkit and a collection of scripts and components intended to accelerate development of HoloLens application development, and extends it by adding an holographic user interface.

Chapter 6 , *Interacting with Holograms Using Unity*, explores various ways the user can interact with holograms. We start by discussing gestures, along with allowing the user to *touch* holograms, then introduce the reader to Voice User Interfaces, walking the reader through extending the application to allow the user to manipulate the hologram using their voice.

Chapter 7, *Collaboration with HoloLens Using Unity*, outlines building a project where HoloLens is shared across multiple devices. The example is based on a collaborative design review, whereby a design (a 3D model) is streamed to multiple HoloLens devices where it is projected into the real world, allowing for richer engagement and collaboration.

Chapter 8, *Developing a Multiplayer Game Using Unity*, covers our final example of building a fun multiplayer game inspired by the classic Cannon Fodder, where the user will navigate their team around the environment to eliminate their opponent.

Chapter 9, *Deploying Your Apps to the HoloLens Device and Emulator*, showcases the process of building and deploying applications to an actual HoloLens device and/or emulator.

What you need for this book

Here is a list of software used in this book:

- Window 10 Pro
- Visual Studio 2015 Update 3
- HoloLens Emulator (build 10.0.14393.0)
- Unity 5.5.0f3
- Holotoolkit-Unity (SHA1 cd2fd2f9569552c37ab9defba108e7b7d9999b12)
- Vuforia 6.1
- Blender 2.78

Who this book is for

This book is for developers who have some experience with programming in any of the major languages, such as C# and C++. You do not need any knowledge of augmented reality development.

Conventions

In this book, you will find a number of text styles that distinguish between different kinds of information. Here are some examples of these styles and an explanation of their meaning. Code words in text, database table names, folder names, filenames, file extensions, pathnames, dummy URLs, user input, and Twitter handles are shown as follows: "The details of spatial reasoning are encapsulated in the `HolographicFrame` class, which is created at each frame within the `Update` method, as in the following extract."

A block of code is set as follows:

```
public HolographicFrame Update()
{
// ...
HolographicFrame holographicFrame =
holographicSpace.CreateNextFrame();
// ...
}
```

When we wish to draw your attention to a particular part of a code block, the relevant lines or items are set in bold:

```
public void SetHolographicSpace(HolographicSpace holographicSpace)
{
// ...
referenceFrame =
locator.CreateStationaryFrameOfReferenceAtCurrentLocation();
}
```

New terms and **important words** are shown in bold. Words that you see on the screen, for example, in menus or dialog boxes, appear in the text like this: "Next, launch Visual Studio, and create a new project by selecting the **File | New Project** menu item."

Warnings or important notes appear like this.

Tips and tricks appear like this.

Reader feedback

Feedback from our readers is always welcome. Let us know what you think about this book-what you liked or disliked. Reader feedback is important for us as it helps us develop titles that you will really get the most out of. To send us general feedback, simply email feedback@packtpub.com, and mention the book's title in the subject of your message. If there is a topic that you have expertise in and you are interested in either writing or contributing to a book, see our author guide at www.packtpub.com/authors.

Customer support

Now that you are the proud owner of a Packt book, we have a number of things to help you to get the most from your purchase.

Downloading the example code

You can download the example code files for this book from your account at http://www.packtpub.com. If you purchased this book elsewhere, you can visit http://www.packtpub.com/support and register to have the files emailed directly to you. You can download the code files by following these steps:

1. Log in or register to our website using your email address and password.
2. Hover the mouse pointer on the **SUPPORT** tab at the top.
3. Click on **Code Downloads & Errata**.
4. Enter the name of the book in the **Search** box.
5. Select the book for which you're looking to download the code files.
6. Choose from the drop-down menu where you purchased this book from.
7. Click on **Code Download**.

Once the file is downloaded, please make sure that you unzip or extract the folder using the latest version of:

- WinRAR / 7-Zip for Windows
- Zipeg / iZip / UnRarX for Mac
- 7-Zip / PeaZip for Linux

The code bundle for the book is also hosted on GitHub at
`https://github.com/PacktPublishing/Microsoft-HoloLens-By-Example`. We also have
other code bundles from our rich catalog of books and videos available at
`https://github.com/PacktPublishing/`. Check them out!

Downloading the color images of this book

We also provide you with a PDF file that has color images of the screenshots/diagrams used
in this book. The color images will help you better understand the changes in the output.
You can download this file from
`https://www.packtpub.com/sites/default/files/downloads/MicrosoftHoloLensByExamp`
`le_ColorImages.pdf`.

Errata

Although we have taken every care to ensure the accuracy of our content, mistakes do
happen. If you find a mistake in one of our books-maybe a mistake in the text or the code-
we would be grateful if you could report this to us. By doing so, you can save other readers
from frustration and help us improve subsequent versions of this book. If you find any
errata, please report them by visiting `http://www.packtpub.com/submit-errata`, selecting
your book, clicking on the **Errata Submission Form** link, and entering the details of your
errata. Once your errata are verified, your submission will be accepted and the errata will
be uploaded to our website or added to any list of existing errata under the Errata section of
that title. To view the previously submitted errata, go to
`https://www.packtpub.com/books/content/support` and enter the name of the book in the
search field. The required information will appear under the **Errata** section.

Piracy

Piracy of copyrighted material on the internet is an ongoing problem across all media. At
Packt, we take the protection of our copyright and licenses very seriously. If you come
across any illegal copies of our works in any form on the internet, please provide us with
the location address or website name immediately so that we can pursue a remedy. Please
contact us at `copyright@packtpub.com` with a link to the suspected pirated material. We
appreciate your help in protecting our authors and our ability to bring you valuable
content.

Questions

If you have a problem with any aspect of this book, you can contact us at questions@packtpub.com, and we will do our best to address the problem.

1
Enhancing Reality

Welcome to Microsoft HoloLens By Example; join me on a journey through this book to learn about Microsoft's HoloLens device and more generally, **mixed reality** (**MR**) applications, through a series of exciting example, each uncovering an important concept related to developing mixed reality applications, including:

- Understanding the environment and context of the user through image recognition and spatial mapping
- Projecting and placing holograms into the real world
- Allowing the user to interact with holograms using a variety interaction modes including gaze, gesture and voice
- Sharing the experience across devices

We will start our journey by briefly peering into the past to see how our relationship with computers has changed over the past couple of decades, before defining MR with respect to reality and MR. We will then cover some of the core building blocks of MR applications before wrapping up this chapter.

A good place to start is always at the beginning, so let's start there.

A brief look at the past

Let's start our journey by briefly peering into the past, specifically at how our interaction with computers has changed over time and how it might be in the near future.

Early computers, bypassing when people were used as computers, were large mechanical machines, and their size and cost meant that they were fixed to a single location and limited to a specific task. The limitation of a single function was soon resolved with **Electronic Numerical Integrator And Computer (ENIAC)**, one of the world's first general-purpose electronic computers. Due to its size and cost, it was still fixed to a single location and interacted with/programmed through a complex process of rearranging physical switches on a switch board. Mainframes followed and introduced a new form of interaction, the **Command Line Interface (CLI)**. The user would interact with the computer by issuing a series of instructions and have the response returned to them via a Terminal screen. Once again, these computers were expensive, large, and complex to use.

The following screenshot is an example of DOS, an operating system that dominated the personal computing market in the late 1980s:

```
Current date is Tue  1-01-1980
Enter new date:
Current time is  7:48:27.13
Enter new time:


The IBM Personal Computer DOS
Version 1.10 (C)Copyright IBM Corp 1981, 1982

A>dir/w
COMMAND  COM     FORMAT   COM     CHKDSK   COM     SYS      COM     DISKCOPY COM
DISKCOMP COM     COMP     COM     EXE2BIN  EXE     MODE     COM     EDLIN    COM
DEBUG    COM     LINK     EXE     BASIC    COM     BASICA   COM     ART      BAS
SAMPLES  BAS     MORTGAGE BAS     COLORBAR BAS     CALENDAR BAS     MUSIC    BAS
DONKEY   BAS     CIRCLE   BAS     PIECHART BAS     SPACE    BAS     BALL     BAS
COMM     BAS
        26 File(s)
A>dir command .com
COMMAND  COM      4959   5-07-82  12:00p
         1 File(s)
A>
```

PC DOS 1.10 screenshot, Credit: Leyo, Source: https://commons.wikimedia.org/wiki/File:PC_DOS_1.10_screenshot.png

The era of direct manipulation interfaces and personal computing followed. With the reduced size, cost, and introduction of the **Graphical User Interface (GUI)**, computers had finally become more accessible to the ordinary person. During this era, the internet was born, computers became platforms allowing people to connect and collaborate with one another, to be entertained, and to augment their skills by making data and information more accessible. However, these computers were, despite their name, still far from being personal, neglecting the user and their current context. These computers forced the user to work in a digital representation of their world, fixing them to a single location--the desk.

The following photograph shows the Apple Macintosh, seen as one of the innovators of the GUIs:

Apple LISA II Macintosh-XL - Credit: Gerhard GeWalt Walter, Source: https://commons.wikimedia.org/wiki/File:Apple-LISA-Macintosh-XL.jpg

BlackBerry Limited (then known as Research In Motion), released its first smartphone in around 2000, which edged us toward the mobile computer. The tipping point was in 2007, when Steve Jobs revealed the iPhone. Smartphones became ubiquitous; personal computing had finally arrived, and it was a platform that was inherently personal. Constraints of technology provided the catalyst to rethink what the role of the application was. Since then, we have been incrementally improving on this paradigm with more capable devices and smarter services. For a lot of us, our smartphone is our primary device-being always-on and always-connected has given us superpowers our ancestors had only dreamed about.

The following is a photograph showing the late Steve Jobs presenting the first version of the iPhone to the world:

Steve Jobs shows off the iPhone 4 at the 2010 Worldwide Developers Conference - Credit: Matthew Yohe.
Source: https://en.wikipedia.org/w/index.php?title=File:Steve_Jobs_Headshot_2010.JPG

Taking a bird's-eye view, we can see the general (and obvious) trend of the following things:

- Minimization of hardware
- Increase in utility
- Moving toward more natural ways of interacting
- The shift from us being in the computer world toward computers being in our world

So, what does the next paradigm look like? I believe that HoloLens gives us a glimpse into what computers will look like in the near future, and even in its infancy, I believe that it provides us with a platform to start exploring and defining what the future will look like. Let's continue by clarifying the differences and similarities of **virtual reality** (**VR**), **augmented reality** (**AR**), and MR.

The following is a photograph illustrating an example of how MR seamlessly blends holograms into the real world:

Win10 HoloLens Minecraft, Credit: Microsoft Sweden, Source: https://www.flickr.com/photos/microsoftsweden/15716942894

Defining mixed reality

Let's begin by first defining and contrasting three similar, but frequently misused, paradigms: VR, AR, and MR:

- VR describes technology and experiences where the user is fully immersed in a virtual environment.
- AR can be described as technology and techniques used to superimpose digital content onto the real world.
- MR can be considered as a blend of VR and AR. It uses the physical environment to add realism to holograms, which may or may not have any physical reference point (as AR does).

The differences between VR, AR, and MR are not so much in the technology but in the experience you are trying to create. Let's illustrate it through an example--imagine that you were given a brief to help manage children's anxiety when requiring hospital treatment.

With VR, you might create a story or game in space, with likeable characters that represent the staff of the hospital in role and character, but in a more interesting and engaging form. This experience will gently introduce the patient to the concepts, procedures, and routines required. On the other end of the spectrum, you can use augmented reality to deliver fun facts based on contextual triggers, for example, the child might glance (with their phone or glasses) at the medicine packaging to discover what famous people had their condition. MR, as the name suggests, mixes both the approaches--our solution can involve a friend, such as a teddy, for the child to converse with, expressing their concerns and fears. Their friend will accompany them at home and in the hospital, being contextually sensitive and respond appropriately.

As highlighted through these hypothetical examples, they are not mutually exclusive but adjust the degree to which they preserve reality; this spectrum is termed as the **reality–virtuality continuum,** coined by Paul Milgram in his paper *Augmented Reality: A class of displays on the reality-virtuality continuum, 2007.* He illustrated it graphically, showing the spectrum of reality between the extremes of real and virtual, showing how MR encompasses both AR and **augmented virtuality** (**AV**). The following is a figure by Paul Milgram and Fumio Kishino that defined the concept of Milgram's reality-virtuality continuum and illustrates the concept--to the far left, you have reality and at the opposite end, the far right, you have virtuality--as MR strides itself between these two paradigms:

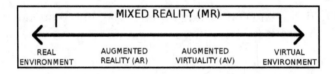

Representation of reality-virtuality continuum by Paul Milgram and Fumio Kishino

Our focus in this book, and the proposition of HoloLens, is MR (also referred to as Holographic). Next, we will look into the principles and building blocks that make up MR experiences.

Designing for mixed reality

The emergence of new technologies is always faced with the question of *doing old things in a new way* or *doing new things in new ways*. When the TV was first introduced, the early programs were adopted from radio, where the presenter read in front of a camera, neglecting the visual element of the medium. A similar phenomenon happened with computers, the web, and mobile--I would encourage you to think about the purpose of what you're trying to achieve rather than the process of how it is currently achieved to free you to create new and innovative solutions.

In this section, we will go over some basic design principles related to building MR experiences and the accompanying building blocks available on HoloLens. Keeping in mind that this medium is still in its infancy, the following principles are still a work in progress.

Identifying the type of experience you want to create

As discussed earlier, the degree of reality you want to preserve is up to you--the application designer. It is important to establish where your experience fits early on as it will impact how you design and also the implementation of the experience you are building. Microsoft outlines three types of experiences:

- **Enhanced environment apps**: These are applications that respect the real world and supplement it with holographic content. An example of this can be pinning a weather poster near the front door, ensuring that you don't forget your umbrella when the forecast is for rain.

- **Blended environment apps**: These applications are aware of the environment, but will replace parts of it with virtual content. An application that lets the user replace fittings and furniture is an example.

- **Virtual environment apps**: These types of applications will disregard the environment and replace it completely with a virtual alternative. An application that converts your room into a jungle, with trees and bushes replacing the walls and the floor can be taken as an example.

Like with so many things, there is no right answer, just a good answer for a specific user, specific context, and at a specific time. For example, designing a weather app for a professional might have the weather forecast pinned to the door so that she sees it just before leaving for work, while it might be more useful to present the information through a holographic rain cloud, for example, to a younger audience.

In the next section, we will continue our discussion on the concepts of MR, specifically looking at how HoloLens makes sense of the environment.

Understanding the environment

One of the most compelling features of HoloLens is its ability to place and track virtual/digital content in the real world. It does this using a process known as **spatial mapping**, whereby the device actively scans the environment, building its digital representation in memory. In addition, it adds *anchors* using a concept called **spatial anchors.** Spatial anchors mark important points in the world in reference to the defined world origin; holograms are positioned relative to these spatial anchors, and these anchors are also used to join multiple spaces for handling larger environments.

The effectiveness of the scanning process will determine the quality of the experience; therefore, it is important to understand this process in order to create an experience that effectively captures sufficient data about the environment. One technique commonly used is **digital painting**; during this phase, the user is asked to paint the environment. As the user glances around, the scanned surfaces are visualized (or painted over), providing feedback to the user that the surface has been scanned.

However, scanning and capturing the environment is just one part of understanding the environment, and the second is making use of it; some uses include the following:

- **Occlusion**: One of the shortfalls of creating immersive MR experiences using single camera devices (such as Smartphones) is the inability to understand the surface to occlude virtual content from the real world when obstructed. Seeing holograms through objects is a quick way to force the user out of the illusion; with HoloLens, occluding holograms with the real world is easy.
- **Visualization**: Sometimes, visualizing the scanned surfaces is desirable, normally an internal effect such as feeding back what part of the environment is scanned to the user.
- **Placement**: Similar to occlusion in that it creates a compelling illusion, holograms should behave like the real objects that they are impersonating. Once the environment is scanned, further processing can be performed to gain greater knowledge of the environment, such as the types of surfaces available. With this knowledge, we can better infer where objects belong and how they should behave. In addition to creating more compelling illusions, matching the experience with the user's mental model of where things belong makes the experience more familiar, thus easing adoption by making it easier and more intuitive to use.

- **Physics**: HoloLens makes the scanned surfaces accessible as plain geometry data, which means we can leverage the existing physics simulation software to reinforce the presence of holograms in the user's environment. For example, if I throw a virtual ball, I expect it to bounce off the walls and onto the floor before settling down.
- **Navigation**: In game development, we have devised effective methods for path planning. Having a digital representation of our real world affords us to utilize these same techniques in the real world. Imagine offering a visually impaired person an opportunity to effectively navigate an environment independently or assisting a parent to find their lost child in a busy store.
- **Recognition**: Recognition refers to the ability of the computer to classify what objects are in the environment; this can be used to create a more immersive experience, such as having virtual characters sit on seats, or to provide a utility, such as helping teach a new language or assisting visually impaired people so that they can better understand their environment.

Thinking in terms of the real world

The luxury of designing for screen-based experiences is that your problem is simplified. In most cases, we own the screen and have a good understanding of it; we lose these luxuries with MR experiences, but gain more in terms of flexibility and therefore opportunity for new, innovative experiences. So it becomes even more important to understand your users and in what context they will be using your application, such as the following:

- Will they be sitting or standing?
- Will they be moving or stationary?
- Is the experience time dependent?

Some common practices when embedding holograms in the real world include the following:

- Place holograms in convenient places--places that are intuitive, easily discovered, and in reach, especially if they are interactive.
- Design for the constraints of the platform, but keep in mind that we are developing for a platform that will rapidly advance in the next few years. At the time of writing, Microsoft recommends placing holograms between 1.25 meters and 5 meters away from the device, with the optimum viewing distance of 2 meters. Find ways of gracefully fading content in and out when it gets too close or far, so as not to jar the user into an unexpected experience.

- As mentioned earlier, placing holograms on contextually relevant surfaces and using shadows create, more immersive experiences, giving a better illusion that the hologram exists in the real world.
- Avoid locking content to the camera; this can quickly become an annoyance to the user. Rather, use an alternative that is more gentle, an approach being adopted has the interface dragged, in an elastic-like manner, with the user's gaze.
- Make use of **spatial sound** to improve immersion and assist in hologram discovery. If you have ever listened to Virtual Barber Shop Hair Cut (`https://www.youtube.com/watch?v=8IXm6SuUigI`), you will appreciate how effective 3D sound can be in creating an immersive experience and, similar to mimicking the behavior of the objects you are trying to impersonate, use real world sound that the user will expect from the hologram.

The spatial sound, such as 3D, adds another dimension to how sound is perceived. Sounds are normally played back in stereo, meaning that the sound has no spatial position, that is, the user won't be able to infer where in space the sound comes from. Spatial sound is a set of techniques that mimic sound in the real world. This has many advantages, from offering more realism in your experience to assisting the user locate content.

Of course, this list is not comprehensive, but has a few practices to consider when building MR applications. Next, we will look at ways in which the user can interact with holograms.

Interacting in mixed reality

With the introduction of any new computing paradigm comes new ways of interacting with it and, as highlighted in the opening paragraph, history has shown that we are moving from an interface that is natural to the computer toward an interface that is more natural to people. For the most part, HoloLens removes dedicated input devices and relies on inferred intent, gestures, and voice. I would argue that this constraint is the second most compelling offering that HoloLens gives us; it is an opportunity to invent more natural and seamless experiences that can be accessible to everyone. Microsoft refers to three main forms of input, including **Gaze Gesture Voice** (**GGV**); let's examine each of these in turn.

Gaze refers to tracking what the user is looking at; from this, we can infer their interest (and intent). For example, I will normally look at a person before I speak to them, hopefully, signalling that I have something to say to them. Similarly, during the conversation, I may gaze at an object, signalling to the other person that the object that I'm gazing at is the subject I'm speaking about.

This concept is heavily used in HoloLens applications for selecting and interacting with holograms. Gaze is accompanied with a cursor; the cursor provides a visual representation of the users gaze, providing visual feedback to what the user is looking at. It can additionally be used to show the state of the application or object the user is currently gazing at, for example, the cursor can visually change to signal whether the hologram the user is gazing at is interactive or not. On the official developer site, Microsoft has listed the design principles; I have paraphrased and listed them here for convenience:

- **Always present**: The cursor is, in some sense, akin to the mouse pointer of a GUI; it helps the users understand the environment and the current state of the application.

- **Cursor scale**: As the cursor is used for selecting and interacting with holograms, it's size should be no bigger than the objects the user can interact with. Scale can also be used to assist the users' understanding of depth, for example, the cursor will be larger when on nearby surfaces than when on surfaces farther away.

- **Look and feel**: Using a directionless shape means that you avoid implying any specific direction with the cursor; the shape commonly used is a donut or torus. Making the cursor hug the surfaces gives the user a sense that the system is aware of their surroundings.

- **Visual cues**: As mentioned earlier, the cursor is a great way of communicating to the user about what is important as well as relaying the current state of the application. In addition to signalling to the user what is interactive and what is not, it also can be used to present additional information (possible actions) or the current state, such as visualizing showing the user that their hand has been detected.

While gazing provides the mechanism for targeting objects, gestures and voice provide the means to interact with them. Gestures can be either discrete or continuous. The discrete gestures execute a specific action, for example, the air-tap gesture is equivalent to a double-click on a mouse or tap on the screen. In contrast, continuous gestures are entered and exited and while active, they will provide continuous update to their state. An example of this is the manipulation gesture, whereby the user enters the gesture by holding their finger down (called the hold gesture); once active, this will continuously provide updates of the position of the tracked hand until the gesture is exited with the finger being lifted. This is equivalent to dragging items on desktop and touch devices with the addition of depth.

HoloLens recognizes and tracks hands in either the ready state (back of hand facing you with the index finger up) or pressed state (back of hand facing you with the index finger down) and makes the current position and state of the currently tracked hands available, allowing you to devise your own gestures in addition of providing some standard gestures, some of which are reserved for the operating system. The following gestures are available:

- **Air-tap**: This is when the user presses (finger down) and releases (finger up), and is performed within a certain threshold. This interaction is commonly associated to selecting holograms (as mentioned earlier).

- **Bloom**: Reversed for the operating system, bloom is performed by holding your hand in front of you with your fingers closed, and then opening your hand up. When detected, HoloLens will redirect the user to the **Start** menu.

- **Manipulation**: As mentioned earlier, manipulation is a continuous gesture entered when the user presses their finger down and holds it down, and exited when hand tracking is lost or the user releases their finger. When active, the user's hand is tracked with the intention of using the absolute position to manipulate the targeted hologram.

- **Navigation**: This is similar to the manipulation gesture, except for its intended use. Instead of mapping the absolute position changes of the user's hand with the hologram, as with manipulation, navigation provides a standard range of -1 to 1 on each axis (x, y, and z); this is useful (and often used) when interacting with user interfaces, such as scrolling or panning.

The last dominate form of interacting with HoloLens, and one I'm particularly excited about, is voice. In the recent times, we have seen the rise of **Conversational User Interface** (**CUI**); so, it's timely to introduce a platform where one of it's dominate inputs is voice. In addition to being a vision we have had since before the advent of computers, it also provides the following benefits:

- Hands free (obviously important for a device like HoloLens)
- More efficient and requires less effort to achieve a task; this is true for data entry and navigating deeply nested menus
- Reduces cognitive load; when done well, it should be intuitive and natural, with minimal learning required

However, how voice is used is really dependent on your application; it can simply be used to supplement gestures such as allowing the user to use the `Select` keyword (a reserved keyword) to select the object the user is currently gazing at or support complex requests by the user, such as answering free-form questions from the user. Voice also has some weaknesses, including these:

- Difficulty with handling ambiguity in language; for example, how do you handle the request of `louder`
- Manipulating things in physical space is also cumbersome
- Social acceptance and privacy are also considerations that need to be taken into account

With the success of Machine Learning (**ML**) and adoption of services such as Amazon's *Echo*, it is likely that these weaknesses will be short lived.

Summary

So far, we have discussed a lot of high-level concepts, let's now wrap this chapter up before moving on and putting these concepts into practice through a series of examples.

I hope you're as excited as I am and, with this book, join me in shaping the future. This book consists of a series of examples, each walking through a "toy" example used to demonstrate a specific concept or feature of the HoloLens. As you work your way through this book, I encourage you to dream of what is possible, looking past some of the current nuances, knowing that they will be resolved in the near future. I would also discourage creating horseless carriages; a phrase used by the notable designer Don Norman in reference to how the car was designed (and named) on horse-drawn carriages, highlighting how new technology is always started by making it look like the old technology. So, rather than adapting from the existing apps, be inspired to adapt from the real world--with that said, let's make a start with our first example.

2
Tagging the World Using DirectX

One of the most significant advancements of HoloLens has little to do with the actual technology and more to do with where it is placed. Having a computer situated conveniently on your head lends itself to many interesting avenues. It means having a computer easily accessible any place and in any situation and, equally important, having a computer that is capable of seeing what you see and hearing what you hear. Smartphones have taken us to a paradigm where we have had always-on and always-connected computers, but this extends this concept even further, whereby our devices can be more aware of our situation, but of course this is only possible if the applications we build for these platforms take advantage of these opportunities. In this chapter, we start exploring this concept by way of example, whereby we will use Microsoft's Cognitive Services (`Microsoft's Cognitive Services`), to recognize people in the user's view and, if recognized, display associated metadata--name in this case. While reading through this chapter, I encourage you to let your mind drift and imagine how else augmentation can be applied to enhance people's lives.

While working through this chapter (and example), we will touch on the following topics:

- A brief walkthrough of the Visual C# Holographic DirectX 11 App template provided by Microsoft
- Capturing frames from the HoloLens color camera
- Using Microsoft's Cognitive Services to detect and identify faces
- Positioning a name tag relative to a recognized face in the users view

Creating a mixed reality app

Microsoft defines an MR application as an application developed for the Windows 10 Universal Platform, utilizing the holographic rendering, gaze, gesture, motion, and voice APIs. In this chapter, and in the next, we will focus on building MR applications using DirectX 11, C#, and SharpDX (C# wrapper for DirectX 11); alternatively, you can follow along using C++.

 DirectX is a large topic (and one of many); it is not the intention of this book to cover DirectX, but to use it as a platform to illustrate the examples presented in this book. For a more comprehensive look at DirectX, for C#, I recommend *Direct 3D Rendering Cookbook* by Justin Stenning (Direct3D Rendering Cookbook).

We will start by walking through the Visual C# Holographic DirectX 11 App Template, focusing on sections related to HoloLens, before moving on to extend it for the example of this chapter.

To begin with, ensure that you have all the dependencies installed; details of these can be found on the official Microsoft at https://developer.microsoft.com/en-us/windows/mixed-reality/install_the_tools.

This book uses the following:

- Visual Studio 2015 Update 3 Community
- HoloLens Emulator (build 10.0.14393.0)
- SharpDX3.0.0-beta01 (the latest version at the time of writing this chapter)
- Unity 5.5.0f3 (used in further chapters)
- Vuforia 6.1 (used in a further chapter)

Next, launch Visual Studio, create a new project by selecting the **File** | **New Project** menu item. With the **New Project** dialog open, expand **Templates** | **Visual C#** | **Windows** | **Universal** | **Holographic** and select the **Holographic DirectX 11 App (Universal Windows) Template**, as shown in the following screenshot:

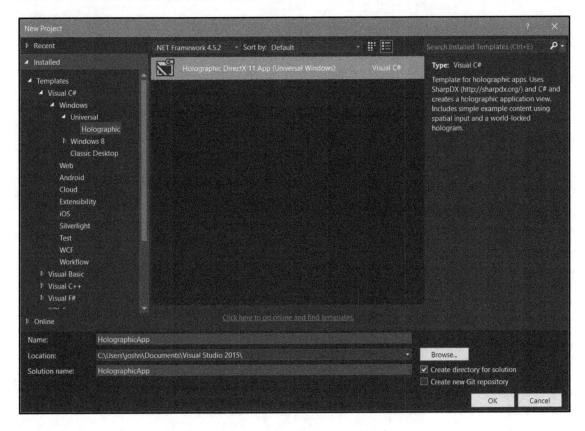

The project template shows how to render a rotating cube at a world-locked location placed 2 meters in front of the user. The user can reposition the cube by performing an air-tap gesture; this will reposition the cube 2 meters in front of the user's gaze. Along with providing best practices when developing for the HoloLens, the template contains a set of helpful utility classes, bundled in the Common namespace, that can be taken advantage of in your own projects.

With the project now created, let's explore some of the components unique to HoloLens, starting with the class HolographicSpace. When the class CoreWindow is (re)created, a corresponding HolographicSpace is instantiated, which is responsible for fullscreen rendering, camera data, and access to spatial reasoning APIs.

The `HolographicSpace` is created within the `SetWindow(CoreWindow window)` method of the `AppView` class, as shown in the following code snippet:

```
public void SetWindow(CoreWindow window)
{
// ...
holographicSpace = HolographicSpace.CreateForCoreWindow(window);

deviceResources.SetHolographicSpace(holographicSpace);

main.SetHolographicSpace(holographicSpace);
}
```

 The official documentation describes the `HolographicSpace` as a representation of a holographic scene, with one or more holographic cameras rendering its content. It essentially controls the rendering pipeline for holographic applications.

Once the `HolographicSpace` is instantiated, it is bound to the `deviceResource`, which in turn, uses it to create the DXGI adapter, an API used to access the video hardware. The `HolographicSpace` is also passed to the main application, `HolographicAppMain`, where the application registers for the `CameraAdded` and `CameraRemoved` events. The following snippet shows an extract of the `SetHolographicSpace` method of the `HolographicAppMain`:

```
public void SetHolographicSpace(HolographicSpace holographicSpace)
{
this.holographicSpace = holographicSpace;

// ...
holographicSpace.CameraAdded += this.OnCameraAdded;
holographicSpace.CameraRemoved += this.OnCameraRemoved;
}
```

Device-based resources need to be created or disposed accordingly for each camera; this is the responsibility of the application and is performed within the `CameraAdded` and `CameraRemoved` events, respectively.

The other significant role the `HolographicSpace` plays is in spatial reasoning. As the name suggests, **spatial reasoning** is concerned with predicting the specific location of the user's head (also known as device) for a given time, which of course, are essentially our cameras (World View Matrices) and is used to accurately position the holograms. The details of spatial reasoning are encapsulated in the `HolographicFrame` class, which is created at each frame within the `Update` method, as in the following extract:

```
public HolographicFrame Update()
{
// ...
HolographicFrame holographicFrame =
holographicSpace.CreateNextFrame();
// ...
}
```

For those who are unfamiliar with the use of matrices in 3D programming, matrices are essentially used to transform points (vertices) into different spaces. In 3D, we are normally dealing with the matrices model, world, and view. The **matrix model** is specifically for a single object, for example, cube. It determines where it resides in space, encapsulating (as with all transformation matrices) position, rotation, and scale. The **world** matrix defines where and how the object is placed inside the scene. Finally, the **view** (or camera) matrix is position relative to the camera.

They are used extensively in 3D programming, and in this book, so if you are unfamiliar with these concepts, I strongly recommend reading some material on 3D Math for game programming, such as *Mathematics for 3D Game Programming and Computer Graphics*, by Eric Lengyel.

The `HolographicFrame` exposes two objects that are used to better understand the user's current position and pose, the `HolographicFramePrediction` and `SpatialCoordinateSystem`; let's take a look at each of these in turn.

The `HolographicFramePrediction` encapsulates the camera properties and pose. The camera properties are used to (re)create the associated buffers, while the pose is used to update the view and projection matrices per frame.

The other object made available by the `HolographicFrame` is the `SpatialCoordinateSystem`. HoloLens provides a unique approach in managing coordinate systems, a concept we frequently revisit throughout this book. It essentially, unlike virtual environments that consist of a single coordinate system, manages multiple spaces, and does so by creating multiple coordinate systems for each space which, in turn, are used by the device to reason about the position and orientation of the holograms. The following figure illustrates this concept, where the VR environment consists of a single frame of reference, while the MR consists of one or more:

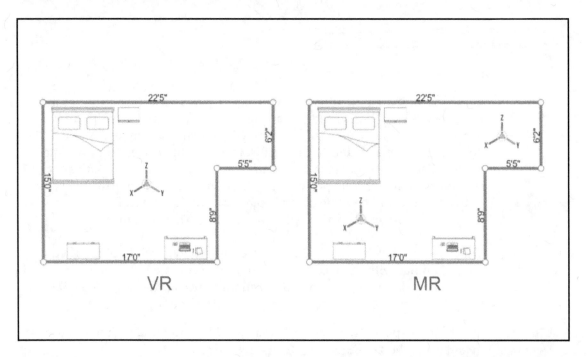

The following code extract shows an extended version of the preceding `Update` method to include references of `HolographicFramePrediciton` and `SpatialCoordinateSystem`, showing snippets of how the `HolographicFramePrediction` is being used:

```
public HolographicFrame Update()
    {
    HolographicFrame holographicFrame =
    holographicSpace.CreateNextFrame();

    HolographicFramePrediction prediction =
    holographicFrame.CurrentPrediction;

    deviceResources.EnsureCameraResources(holographicFrame, prediction);
```

```
SpatialCoordinateSystem currentCoordinateSystem =
referenceFrame.CoordinateSystem;

// ...

SpatialInteractionSourceState pointerState =
spatialInputHandler.CheckForInput();
if (null != pointerState)
{
spinningCubeRenderer.PositionHologram(
pointerState.TryGetPointerPose(currentCoordinateSystem)
);
}

timer.Tick(() =>
{

// ...

foreach (var cameraPose in prediction.CameraPoses)
{
HolographicCameraRenderingParameters renderingParameters =
holographicFrame.GetRenderingParameters(cameraPose);

renderingParameters.SetFocusPoint(
currentCoordinateSystem,
spinningCubeRenderer.Position
);
}
return holographicFrame;
}
```

 It is not the intention of this chapter to go into any depth of the template but rather provide a top level tour of where and how the major components are used. I encourage you to read Microsoft's official document on the HoloLens developer site and examine the code.

Similarly, the following is an extract of the Render method, showing snippets of how HolographicFramePrediction is being used:

```
public bool Render(ref HolographicFrame holographicFrame)
{

// ...

holographicFrame.UpdateCurrentPrediction();
HolographicFramePrediction prediction =
```

```
holographicFrame.CurrentPrediction;

return deviceResources.UseHolographicCameraResources(
(Dictionary<uint, CameraResources> cameraResourceDictionary) =>
{
bool atLeastOneCameraRendered = false;

foreach (var cameraPose in prediction.CameraPoses)
{
CameraResources cameraResources =
cameraResourceDictionary[cameraPose.HolographicCamera.Id];

// ...

cameraResources.UpdateViewProjectionBuffer(deviceResources,
cameraPose, referenceFrame.CoordinateSystem);

bool cameraActive =
cameraResources.AttachViewProjectionBuffer(deviceResources);

if (cameraActive)
{
spinningCubeRenderer.Render();
}
atLeastOneCameraRendered = true;
}

return atLeastOneCameraRendered;
});
}
```

There is a lot going on, but for the purpose of our discussion, let's narrow our focus just to how `HolographicFramePrediction` and `SpatialCoordinateSystem` are used. Within the `Update` method, the `SpatialCoordinateSystem` is used to determine the user's position and pose, but only when an air-tap gesture is detected. We will return to the air-tap gesture later on in this chapter. Within the `Update` method, we also set the focus point for each camera; this is an optional step that can be used to further improve the quality of tracking by having the HoloLens prioritize areas of stabilization, normally near where you holograms reside.

To ensure that holograms are locked in position, HoloLens performs image stabilization that will attempt to compensate for movement in the display around the focal plane. A single plane, called the **stabilization plane**, is used to maximize stabilization for the holograms in the vicinity. HoloLens will automatically position the stabilization plane, but it can be overridden by the user, as seen earlier. The `SetFocusPoint` has multiple overrides, allowing more granular control over how the stabilization plane behaves, such as the normal and velocity relative to the holograms.

Now, let's turn our attention to the `Render` method to ensure minimal latency between the user and the virtual cameras; `UpdateCurrentPrediction` is called on the `HolographicFrame`, which ensures that it has the most up-to-date frame for rendering.

Next, we iterate over each camera updating the associated buffers (`RenderTargets` and `DepthStencilView`) of the context, including updating the view and projection matrices for a given camera's current pose and coordinate system.

So far, we briefly mentioned how HoloLens differs from purely virtual environments with respect to their coordinate system, without delving into detail about how this coordinate system is established. We will now revisit this and dive a little deeper into how this works. In order to be able to position Holograms, we need to establish a central point of reference such that the position--0,0,0--has the same meaning for all holograms in your environment. This central point of reference comes from a `FrameOfReference` created by the `SpatialLocator`, a class responsible for tracking the motion of HoloLens. If you examine the `SetHolographicSpace` of the `HolographicApp` class, you can see where this happens, as shown in the following code snippet:

```
public void SetHolographicSpace(HolographicSpace holographicSpace)
{
// ...
locator = SpatialLocator.GetDefault();
locator.LocatabilityChanged += this.OnLocatabilityChanged;
}
```

As you can see, you obtain the `SpatialLocator` via the `GetDefault()` class method; with reference to the `SpatialLocator`, we can now register to be notified of tracking state changes, such as if the device loses its ability to track, in which case you won't be able to render the holograms. Using the `SpatialLocator`, we can create a `FrameOfReference`, which acts as the central point of our digital world such that the holograms in proximity will be positioned and orientated relative to this. The following extract shows where this is done:

```
public void SetHolographicSpace(HolographicSpace holographicSpace)
```

```
{
// ...
referenceFrame =
locator.CreateStationaryFrameOfReferenceAtCurrentLocation();
}
```

With the
`SpatialLocator.CreateStationaryFrameOfReferenceAtCurrentLocation()` call,
we create what is essentially a world locked anchor where the origin is the current position
of HoloLens; remember that everything using the associated coordinate system is
positioned and orientated relative to this point.
The `CreateStationaryFrameOfReferenceAtCurrentLocation` has overrides that
allow you to further offset and orientate the reference frame from the device.

The alternative to the `SpatialStationaryFrameOfReference` (shown in the preceding
section) is the `SpatialLocatorAttachedFrameOfReference`, created via the
`CreateAttachedFrameOfReferenceAtCurrentHeading()` method of the
`SpatialLocator` class. Unlike the stationary frame of reference created, the attached frame
of reference, as the name suggests, is attached to the HoloLens device. The implication of
this is that holograms using its coordinate system are positioned relative to the device. It is
worth noting that the frame of reference has a fixed orientation, that is, it's invariant to the
device rotating. In this example, we rely on the stationary frame of reference and explore
the attached frame of reference in Chapter 3, *Assistant Item Finder Using DirectX.*

The next natural question is how we can use this to position holograms, which leads us
nicely to the discussion of spatial coordinate systems. As mentioned earlier; your objects are
positioned relative to a fixed point (position and orientation) in a purely virtual space. Just
like a purely virtual space, HoloLens positions holograms relative to a reference point, but
unlike a purely virtual space, it needs to reason about the holograms' position and pose
relative to multiple reference points (coordinate systems), which are anchored in the real
world. In this example, the coordinate system is used to obtain `SpatialPointerPose` via
`SpatialInteractionSourceState`, as shown in the following code snippet:

```
public HolographicFrame Update()
{
// ...
SpatialCoordinateSystem currentCoordinateSystem =
referenceFrame.CoordinateSystem;

SpatialInteractionSourceState pointerState =
spatialInputHandler.CheckForInput();
if (null != pointerState)
{
spinningCubeRenderer.PositionHologram(
```

```
pointerState.TryGetPointerPose(currentCoordinateSystem)
);
}
}
```

When reference to the `SpatialPointerPose`, we can obtain the device's position and facing direction relative to the specified coordinate system. We can see how this is used within the method `PositionHologram` of the `SpinningCubeRenderer` as shown in the following code snippet:

```
public void PositionHologram(SpatialPointerPose pointerPose)
{
if (null != pointerPose)
{
Vector3 headPosition = pointerPose.Head.Position;
Vector3 headDirection = pointerPose.Head.ForwardDirection;

float distanceFromUser = 2.0f; // meters
Vector3 gazeAtTwoMeters = headPosition + (distanceFromUser *
headDirection);

this.position = gazeAtTwoMeters;
}
}
```

In the preceding code snippet, we can see how the cube is being positioned 2 meters in front of the user's gaze using `Head.Position` and `Head.ForwardDirection` of the `SpatialPointerPose` instance.

The final chunk we will briefly discuss in this section before moving on to the example is the class `SpatialInteractionManager` and how it is being used in this example. Unlike your traditional desktop or laptop, HoloLens is not intended to have an external keyboard or mouse readily available; therefore, it relies heavily on gaze, gestures, and voice for its primary mode of input. These lend themselves well to the multimodal interaction model where multiple sources can be used to achieve the same task, or used together to achieve a particular task. A simple example of this can be selecting a hologram where the hologram can be selected via an air-tap gesture or via the `select` voice command. `SpatialInteractionManager` is the class responsible for managing these different modes of inputs and mapping them to a consistent gesture that can be handled by your application. To make this more concrete, let's peek inside the `SpatialInputHandler` class to see how this example is making use of it:

```
public class SpatialInputHandler
{
private SpatialInteractionManager interactionManager;
```

```
private SpatialInteractionSourceState sourceState;

public SpatialInputHandler()
{
interactionManager = SpatialInteractionManager.GetForCurrentView();
interactionManager.SourcePressed += this.OnSourcePressed;
}

public SpatialInteractionSourceState CheckForInput()
{
SpatialInteractionSourceState sourceState = this.sourceState;
this.sourceState = null;
return sourceState;
}

public void OnSourcePressed(SpatialInteractionManager sender,
SpatialInteractionSourceEventArgs args)
{
sourceState = args.State;
}
}
```

You obtain a reference to the SpatialInteractionManager via the
GetForCurrentView static method. Now with access, you can register to be notified of
specific types of interactions; in this instance, the SourcePressed event is assigned a
listener, which is fired when a source has entered the pressed state. The delegate is passed
an instance of the SpatialInteractionSourceEventArgs class that references
a SpatialInteractionSourceState class, which encapsulates details of the event, such
as the source (hand, voice, and controller) and associated properties.

We have only lightly touched on SpatialInteractionManager, but we will frequently
revisit to discuss and experiment with it throughout this book.

This now concludes our whirlwind tour of the template and sets us up nicely to walk
through the example for this chapter, but before moving on, now is a good time to build
and deploy the template to see the result of all the code we have been examining. When
built and deployed, you should see a colored cube floating 2 meters in front of you, as
illustrated:

Project setup

Just to recap, the example for this chapter is about performing facial recognition and tagging the recognized person's name (or alias) into the scene. In this section, we will walk through setting up the Microsoft Cognitive Service, which will be responsible for facial detection and identification as well as preparing the dataset that we will use to perform facial identification, that is, photos of people you would like the service to recognize.

We will use one of the Cognitive Services Microsoft services, called Face API, from `Face API`. As the name implies, the service is orientated around detecting and identifying faces but also allows for some interesting metadata, such as the following:

- Age
- Gender
- Likelihood of smiling
- Wearing glasses
- Emotion (a measurement of happiness, sadness, and anger, along with many others)
- Presence of facial hair

Before you can make use of the service, you must register (if not done so already) an Azure account and add the service to your portal, details described in the next section.

Setting up Azure

In this section we will describe the details required to set up the Cognitive Service on Azure; head to `https://portal.azure.com/` and sign in with a valid account (or create an account).

Once logged in, you will be taken to your Azure Dashboard; from here, click on the **+Next** button at the top of the left panel; this will expand a menu. Within this menu, search and click on **AI + Cognitive Services**, which will spawn another menu where we can select **Face API**, as shown:

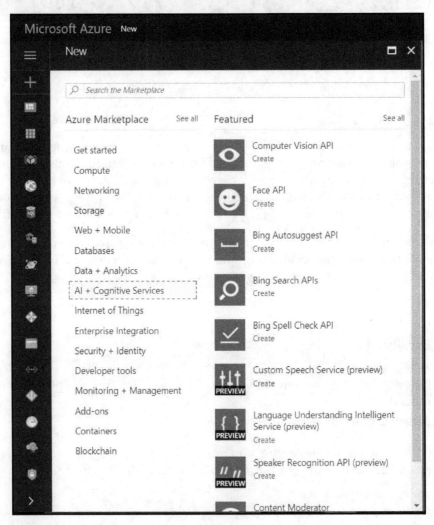

After clicking on the Face API item, you will be presented with a tab requiring you to enter the details of your service, such as Name, Subscription, Location, Pricing tier, and Resource Group. Most of the fields are self-explanatory, but it's important to take note of the location as this affects the URL you use to access the service. With all that done, you have your service up and running; details of this service should be visible on the screen and can be accessed from your Dashboard:

To use your service, you will need to call it with the keys; these can be accessed via the **Show access keys...** link. When clicked on, it will display a set of keys you will need. For convenience, copy and paste them somewhere easily accessible (Notepad) while you read through this chapter.

 Azure offers flexible pricing options with price and limitations dependent on the service and tier selected. At the time of writing, Microsoft offers a free tier for their Face API Cognitive Service. For more details, visit the official site `https://azure.microsoft.com/pricing/details/cognitive-services/face-api` or `https://azure.microsoft.com/pricing` for general pricing.

Creating a group

Before we can recognize a face, we need to teach the service who is who. To do this, we need to create a group, and create people we want to recognize within this group and for each person, upload a set of photos that can be used to train the system. It helps to have a few photos to help the model identify each person. Once all photos have been uploaded, you signal the service to train with the current dataset and only once trained can we use the service to identify the persons we have updated.

Accompanied with this chapter's source code is a Python script to take care of this task; do this now before moving on. Clone or download the source code with this book from the `https://github.com/PacktPublishing/Microsoft-HoloLens-By-Example` repository. Within the `Chapter2/CreateGroup` folder, you will find the `create_group.py` Python script that we can run to take care of this process. Before running this script, we need to build a dataset to upload and train; the script is expecting a folder hierarchy where all images for each person reside in a single directory, with the directory named with the label you would like associated with this person; the hierarchy should resemble the following structure:

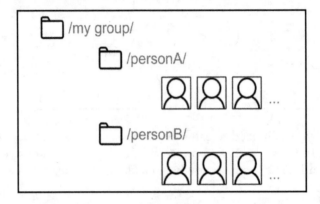

Once created; open up the Command Prompt and navigate to the `create_group.py` script and run, including the parameters shown in the following example:

```
> create_group.py -k <subscription_key> -g <group_id> -d <source_directory>
-o <output_file> [-r <region>]
```

If all goes well, the script will run through the process of creating a group and person for each subdirectory of the root directory provided, train the service, and export a JSON file to be used by your project. This needs to be imported into your project, talking of which, let's now move on to building the example for this project.

The project

It's been a while since we talked about the specifics for the example of this chapter, so let's quickly revise what we are trying to achieve before jumping to the project.

HoloLens includes a world-facing camera color mounted on the front of the device, which enables apps to see what the user sees. The example for this chapter is to create an application that captures what the user sees (the frames from this camera) and uses Microsoft's Cognitive Service Face API that we just set up to detect and identify faces. When a face is recognized, we will display their name along with a predicted age and gender. This example was based on the Microsoft's Holographic face tracking sample, which is available at `https://github.com/Microsoft/Windows-universal-samples/tree/master/Samples/HolographicFaceTracking`.

Let's start by launching the starter solution for this chapter; this can be found in the `Chapter2\Starter` directory from the `https://github.com/PacktPublishing/Microsoft-HoloLens-By-Example` repository. Launch the Visual Studio solution by either double-clicking on the `FaceTag.sln` solution file or by loading from within Visual Studio by clicking on **File** | **Open** | **Project/Solution** and navigating to, and selecting the `FaceTag.sln` file. The starter solution is the standard Visual C# Holographic DirectX 11 App Template we created in the previous section, with the addition of some classes and shaders used to analyze the frame used by the Face API Cognitive service we set up earlier and also used to render the results. Including this in the starter will allow us to focus on HoloLens-specific code, such as capturing the HoloLens frames and positioning the label in the scene (and keep this chapter at a reasonable length).

Adding the Face API key and App Capabilities

If not done so already, copy your services key, which we just described, to your clipboard and paste it into the `cognitive_service.key` file within this project. Without it, the call to the Face API will fail, leaving you with the task of debugging.

Next, we need to add the capabilities of our application to allow us to get access to the internet and camera. Capabilities are a great addition to the Windows platform, giving users more visibility of what applications require, and consequently, more transparency and explicit control of the applications they install. The capabilities are defined in the projects manifest, which can be accessed by opening the **Solution Explorer** panel and double-clicking on the `Package.appxmanifest` file. This will open a new window with the properties of your application; click on the **Capabilities** tab and check the capabilities:

- **Internet (Client)**
- **Webcam**

Your **Capabilities** panel should look similar to the following screenshot with the previously mentioned capabilities checked:

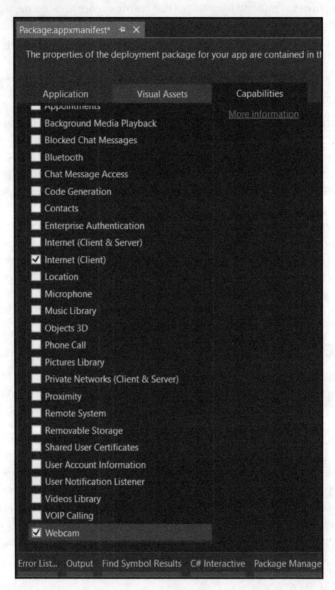

You can learn more about these and the rest of the capabilities at `https://docs.microsoft.com/en-us/windows/uwp/packaging/app-capability-declarations`.

Grabbing frames

In this section, we will create a class that will be responsible for grabbing the frames from the HoloLens color camera and making them available for analysis. Start by creating a new class by right-clicking on the **Content** folder from within the **Solution Explorer** panel, selecting **Add** | **Class**, and then entering the class name `FrameGrabber.cs`.

Accessing the frame will require classes from the `Windows.Media` namespace; to get access to these, add the following at the top of the class:

```
using Windows.Media.Capture;
using Windows.Media.Capture.Frames;
using Windows.Media.Devices.Core;
using Windows.Perception.Spatial;
```

An instance of the class `FrameGrabber` will be created using an asynchronous creation method, allowing the creation of the class without blocking the main thread. Let's define the constructor and then write out the creation method. Make the following amendments to the existing constructor:

```
private FrameGrabber(MediaCapture mediaCapture = null, MediaFrameSource
mediaFrameSource = null, MediaFrameReader mediaFrameReader = null)
{
    this.mediaCapture = mediaCapture;
    this.mediaFrameSource = mediaFrameSource;
    this.mediaFrameReader = mediaFrameReader;

    if (this.mediaFrameReader != null)
    {
        this.mediaFrameReader.FrameArrived +=
        MediaFrameReader_FrameArrived;
    }
}
```

When an instance of the `FrameGrabber` is created, we pass through the created instances of `MediaCapture`, `MediaFrameSource`, and `MediaFrameReader`. Within the constructor, we assign local variables to hold a reference to each of these objects and then register a delegate to handle the `FrameArrived` event--as the name suggests, this is called when frames arrive and we obtain the frames for analysis.

The `MediaCapture` class is used to capture audio, video, and images from a camera, while the `MediaFrameSource` represents the source of media frames, such as a color camera (in this case), and is obtained from the `MediaCapture` instance once it has been initialized. Finally, the `MediaFrameReader` provides access to frames from a `MediaFrameSource` and notifies when a new frame arrives.

Now, let's define the variables and stub out the `MediaFrameReader_FrameArrived` method to remove the errors and warnings; add the following variables to your `FrameGrabber` class:

```
MediaCapture mediaCapture;
MediaFrameSource mediaFrameSource;
MediaFrameReader mediaFrameReader;
```

And similarly let's add the `MediaFrameReader_FrameArrived` method, as in the following code snippet:

```
void MediaFrameReader_FrameArrived(MediaFrameReader sender,
MediaFrameArrivedEventArgs args)
  {
  }
```

With the constructor and variables now defined, let's move on to building the creation method. First, we will define the method along with the local variables; take a look at the following code snippet:

```
public static async Task<FrameGrabber> CreateAsync()
  {
MediaCapture mediaCapture = null;
MediaFrameReader mediaFrameReader = null;

MediaFrameSourceGroup selectedGroup = null;
MediaFrameSourceInfo selectedSourceInfo = null;

  }
```

The next task is to find appropriate `MediaFrameSourceGroup` and `MediaFrameSourceInfo`; these will be used to retrieve the appropriate `MediaFrameSource`. Add the following extract to your `CreateAsync` method:

```
var groups = await MediaFrameSourceGroup.FindAllAsync();
foreach (MediaFrameSourceGroup sourceGroup in groups)
  {
  foreach (MediaFrameSourceInfo sourceInfo in sourceGroup.SourceInfos)
  {
```

```
if (sourceInfo.SourceKind == MediaFrameSourceKind.Color)
{
selectedSourceInfo = sourceInfo;
break;
}
}

if (selectedSourceInfo != null)
{
selectedGroup = sourceGroup;
break;
}
}
```

Here, we are iterating over all the `MeidaFrameSourceGroups`, selecting the first one that has a color source (should only be one for HoloLens). With reference to the `MediaFrameSourceGroup` and `MediaSourceInfo`, we now define the settings--`MediaCaptureInitializationSettings`--and initialize the `MediaCapture`. Append the following to your `CreateAsync` method:

```
var settings = new MediaCaptureInitializationSettings
{
SourceGroup = selectedGroup,
SharingMode = MediaCaptureSharingMode.SharedReadOnly,
StreamingCaptureMode = StreamingCaptureMode.Video,
MemoryPreference = MediaCaptureMemoryPreference.Cpu,
};

mediaCapture = new MediaCapture();

try
{
await mediaCapture.InitializeAsync(settings);
}
catch (Exception e)
{
Debug.WriteLine($"Failed to initilise mediacaptrue {e.ToString()}");
return new FrameGrabber();
}
```

The class `MediaCaptureInitializationSettings` defines the type of `MediaCapture` we want; here, we are initializing the `MediaCapture` to capture the video from a color camera on the CPU. With the settings now defined, we instantiate a new instance of the `MediaCapture` class and try to initialize it via the `InitializeAsync` method.

If initialization is successful, our next task is to obtain the MediaFrameSource and create a MediaFrameReader. The MediaFrameSource can be obtained via the frameSources array of the MediaCapture class, using the selectedSourceInfo.Id as the index lookup. Append the following code to obtain the MediaFrameSource and create the MediaFrameReader:

```
MediaFrameSource selectedSource =
mediaCapture.FrameSources[selectedSourceInfo.Id];

mediaFrameReader = await
mediaCapture.CreateFrameReaderAsync(selectedSource);
```

Finally, we will ensure that the MediaFrameReader has been successfully created. If so, instantiate the FrameGrabber instance passing along the variables we created; otherwise, return an invalid FrameGrabber by passing nothing into the constructor. Append the following extract to the end of your CreateAsync method:

```
MediaFrameReaderStartStatus status = await
mediaFrameReader.StartAsync();

if (status == MediaFrameReaderStartStatus.Success)
{
return new FrameGrabber(mediaCapture, selectedSource,
mediaFrameReader);
}
else
{
return new FrameGrabber();
}
```

With this now implemented; we can create our FrameGrabber, but at this stage it doesn't do anything valuable. We need to obtain reference to the frames for analysis, we will achieve this by holding reference to the latest arrived frame, which can be picked up for analysis. For convenience, we will encapsulate this in a simple data object containing the frame and associated metadata. Add the following class definition within your FrameGrabber class:

```
public struct Frame
{
public MediaFrameReference mediaFrameReference;
public SpatialCoordinateSystem spatialCoordinateSystem;
public CameraIntrinsics cameraIntrinsics;
public long timestamp;
}
```

We will also expose this through a property, along with two more variables for external classes to query the status of the FrameGrabber; one defining whether the FrameGrabber is ready, and the other being the elapsed time since the last frame was captured. Add the following extract to your FrameGrabber class:

```
private Frame _lastFrame;

public Frame LastFrame
{
get
{
lock (this)
{
return _lastFrame;
}
}
private set
{
lock (this)
{
_lastFrame = value;
}
}
}

private DateTime _lastFrameCapturedTimestamp = DateTime.MaxValue;

public float ElapsedTimeSinceLastFrameCaptured
{
get
{
return (float)(DateTime.Now - DateTime.MinValue).TotalMilliseconds;
}
}

public bool IsValid
{
get
{
return mediaFrameReader != null;
}
}
```

As mentioned in the preceding code, the IsValid and ElapsedTimeSinceLastFrameCaptured properties are both used to determine whether the FrameGrabber is ready to avoid having other classes trying to pull frames before the FrameGrabber is ready.

We have two more methods left before we can put our `FrameGrabber` to use; the first is a setter for the Frame, which is a convenience method that will extract the metadata from the captured frame and update the `LastFrame` property. Add the following method to your `FrameGrabber` class:

```
void SetFrame(MediaFrameReference frame)
{
var spatialCoordinateSystem = frame.CoordinateSystem;
var cameraIntrinsics = frame.VideoMediaFrame.CameraIntrinsics;

LastFrame = new Frame
{
mediaFrameReference = frame,
spatialCoordinateSystem = spatialCoordinateSystem,
cameraIntrinsics = cameraIntrinsics,
timestamp = Utils.GetCurrentUnixTimestampMillis()
};

_lastFrameCapturedTimestamp = DateTime.Now;
}
```

When a frame is captured, we pass the frame (instance of `MediaFrameReference`) to this method; it is responsible for updating the `LastFrame` property and timestamp of the `_lastFrameCapturedTimestamp` variable. The `MediaFrameReference` (captured frame) includes the location of the camera as well as the perspective projection of the camera (`CoordinateSystem` and `CameraIntrinsics`) allowing us, developers, to infer the position of the camera in the real world and augment it with digital content.

Finally, let's amend our `MediaFrameReader_FrameArrived` method to pass the arrived frame to our `SetFrame` method defined in the preceding code snippet, as shown:

```
void MediaFrameReader_FrameArrived(MediaFrameReader sender,
MediaFrameArrivedEventArgs args)
{
MediaFrameReference frame = sender.TryAcquireLatestFrame();

if (frame != null && frame.CoordinateSystem != null)
{
SetFrame(frame);
}
}
```

This now concludes our `FrameGrabber` class; now, it's time to put this to use; jump into the application's main class--`FaceTagMain`.

Capturing frames

At the moment, the `FaceTagMain` class is still rendering the cube that we had seen in the first part of this chapter; in this section, we will change it to render the names of the recognized people on labels just in front of them. Let's start by disabling the template by removing the macro `#define DRAW_SAMPLE_CONTENT`. You can do this by simply removing the line or commenting it.

Next, we will set up the `FaceTagMain` class to grab frames using the `FrameGrabber`, which will then be passed to the `FrameAnalyzer` to perform face detection and face identification via the Face API we set up earlier; start by adding the following variable to the `FaceTagMain` class:

```
FrameGrabber frameGrabber;
FrameAnalyzer frameAnalyzer;
```

Next, we will create the `FrameGrabber` and `FrameAnalyzer` during initialization; add the following to the bottom of the `SetHolographicSpace` method:

```
frameGrabber = await FrameGrabber.CreateAsync();
frameAnalyzer = await FrameAnalyzer.CreateAsync();
```

That takes care of the initialization of our `FrameGrabber` and `FrameAnalyzer`; it's time to put them to use. We will declare two methods, of which the first is a convenience method used to test whether our `FrameGrabber` and `FrameAnalyzer` instances are in a valid state to process the next frame; add the following to your `FaceTagMain` class:

```
bool IsInValidateStateToProcessFrame()
{
return
frameGrabber != null && // frameCapture has been initilised
frameGrabber.ElapsedTimeSinceLastFrameCaptured > 0 && // frameCapture has a
frame
frameAnalyzer != null && // frameAnalyzer has been initilised
frameAnalyzer.IsReady; // frameAnalyzer has finished processing the last
frame
}
```

This simply verifies that both the `FrameGrabber` and `FrameAnalzer` have been created and are in a valid state to handle the next frame; details of this are encapsulated in the classes themselves. Next, we will define a method that will take care of processing the frame and calculating where in the physical space any identified faces reside. Add the following method just after `IsInValidateStateToProcessFrame`:

```
void ProcessFrame(SpatialCoordinateSystem worldCoordinateSystem)
{
if (!IsInValidateStateToProcessFrame())
{
return;
}

// obtain the details of the last frame captured
FrameGrabber.Frame frame = frameGrabber.LastFrame;

if (frame.mediaFrameReference == null)
{
return;
}

MediaFrameReference mediaFrameReference = frame.mediaFrameReference;

frameAnalyzer.AnalyzeFrame(frame.mediaFrameReference, (status,
detectedPersons) =>
{
if (status > 0 && detectedPersons.Count > 0)
{

foreach (var dp in detectedPersons)
{
Debug.WriteLine($"Detected person: {dp.ToString()}");
}
}
});

}
```

Here, we are simply verifying that both the `FrameGrabber` and `FrameAnalyzer` are in a valid state; returning from the method early if either of them are not. We then obtain reference to the latest frame and forward the referenced `MediaFrameReference` to the `FrameAnalyzer`, along with a delegate to handle the completed task (successful or not). For now, we simply iterate through all detected faces, outputting their details when the analysis is complete; however, we will return to this later, walk through finding the most relevant face, and update the position of our label (along with the text).

We almost have enough to test; the last thing remaining is to call the `ProcessFrame` method from the `Update` method, as shown in the following code snippet:

```
/// <summary>
/// Updates the application state once per frame.
/// </summary>
public HolographicFrame Update()
{
// ...
SpatialCoordinateSystem currentCoordinateSystem =
referenceFrame.CoordinateSystem;

ProcessFrame(currentCoordinateSystem);
// ...
}
```

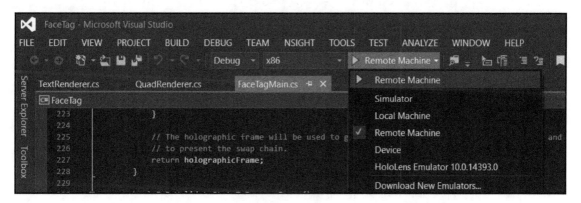

Now is a good time to deploy to the device and test that everything at this stage is working. Click on the **Local Machine** to update the target device. Select **Remote Machine** (as shown in the preceding screenshot), which will display the dialog in the following screenshot. Ensure that HoloLens is *on*, and update the IP address of the **Address** field with the IP address of your HoloLens. Click on **Remote Machine** again to initiate the build and deploy. Once deployed and running, look at one of the persons you uploaded to the Microsoft Face API service to check whether you are successfully capturing and processing the camera frames. If successful, you will see the names of the identified persons printed to the console.

The following figure shows a dialog that will be displayed when first deploying to a remote device:

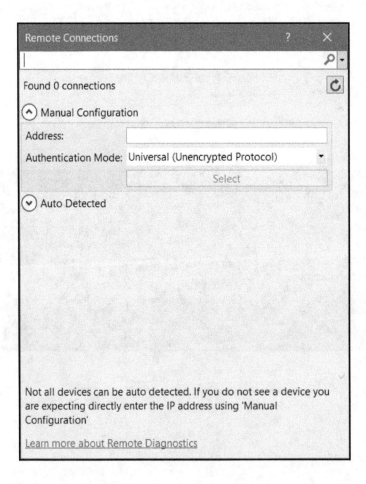

If this is the first time deploying to the HoloLens, it'll ask that you pair the devices. This is just a matter of entering the PIN displayed on the device to the dialog on your computer.

Rendering holographic labels

If you have been following along, your project should now be successfully capturing and analyzing frames. Our next task is to create something to render the results on. We will simply do this using a quad and texture, both of which have been included in the starter project.

Start by declaring the variables to hold references to the classes TextRenderer and QuadRenderer; add the following lines to your FaceTagMain class:

```
TextRenderer textRenderer;
QuadRenderer quadRenderer;
```

The TextRenderer creates and holds references to an in-memory texture; we can write to this texture via the RenderTextOffscreen method, passing the text we would like rendered. The QuadRenderer uses this texture to render. As these are DirectX resources, we are responsible for the creation and disposal of them; make the following amendments to each of the following methods, with the amendments in bold.

In this first snippet, we create the variables textRenderer and quadRenderer; as mentioned earlier, the textRenderer (instance of TextRenderer) is responsible for creating and updating the texture (which will display the recognized persons name/alias), while the quadRenderer (instance of QuardRenderer) is responsible for rendering that texture to a quad in the real world:

```
public void SetHolographicSpace(HolographicSpace holographicSpace)
{
// ...
textRenderer = new TextRenderer(deviceResources, 512, 512);
textRenderer.RenderTextOffscreen("No faces detected");

quadRenderer = new QuadRenderer(deviceResources);
}
```

Also, anytime you create something, we need to dispose of it. In the following code snippet, we add the code responsible for tidying up the resources created by the textRenderer and quadRenderer:

```
public void Dispose()
{
if(textRenderer != null)
{
textRenderer.Dispose();
textRenderer = null;
}

if(quadRenderer != null)
{
quadRenderer.Dispose();
quadRenderer = null;
}
}
```

In the next code snippet, we handle cases for when the device resources are lost and restored. Here, we essentially release any created resources when lost and recreate them once restored:

```
public void OnDeviceLost(Object sender, EventArgs e)
{
textRenderer.ReleaseDeviceDependentResources();
quadRenderer.ReleaseDeviceDependentResources();
}

public void OnDeviceRestored(Object sender, EventArgs e)
{
textRenderer.CreateDeviceDependentResources();
quadRenderer.CreateDeviceDependentResourcesAsync();
}
```

At this stage, we are doing nothing with them; let's rectify this. Jump into the Render method and have it render out QuadRenderer:

```
public bool Render(ref HolographicFrame holographicFrame)
{
if (timer.FrameCount == 0)
{
return false;
}

holographicFrame.UpdateCurrentPrediction();
HolographicFramePrediction prediction =
```

```
holographicFrame.CurrentPrediction;

return deviceResources.UseHolographicCameraResources(
(Dictionary<uint, CameraResources> cameraResourceDictionary) =>
{
bool atLeastOneCameraRendered = false;

foreach (var cameraPose in prediction.CameraPoses)
{
CameraResources cameraResources =
cameraResourceDictionary[cameraPose.HolographicCamera.Id];

var context = deviceResources.D3DDeviceContext;
var renderTargetView = cameraResources.BackBufferRenderTargetView;
var depthStencilView = cameraResources.DepthStencilView;

context.OutputMerger.SetRenderTargets(depthStencilView,
renderTargetView);

SharpDX.Mathematics.Interop.RawColor4 transparent = new
SharpDX.Mathematics.Interop.RawColor4(0.0f, 0.0f, 0.0f, 0.0f);
context.ClearRenderTargetView(renderTargetView, transparent);
context.ClearDepthStencilView(
depthStencilView,
SharpDX.Direct3D11.DepthStencilClearFlags.Depth |
SharpDX.Direct3D11.DepthStencilClearFlags.Stencil,
1.0f,
0);

cameraResources.UpdateViewProjectionBuffer(deviceResources,
cameraPose, referenceFrame.CoordinateSystem);

bool cameraActive =
cameraResources.AttachViewProjectionBuffer(deviceResources);

if (cameraActive)
{
quadRenderer.RenderRGB(textRenderer.Texture);
}
atLeastOneCameraRendered = true;
}

return atLeastOneCameraRendered;
});
}
```

Despite all the code; we have simply added three lines under the `bool cameraActive = cameraResources.AttachViewProjectionBuffer(deviceResources)` statement. So for each active camera, we call `quadRenderer.RenderRGB(textRenderer.Texture)`, which puts it on the list of commands to be rendered.

Our last task is to select and position the label, which we will do next.

Positioning our label

Before jumping into the code, let's review what we will be doing over the next few blocks of code. As we saw earlier in the `FrameGrabbers SetFrame` method; the `MediaFrameReference` contains reference to a `SpatialCoordinateSystem`. As we discussed previously, the `SpatialCoordinateSystem` is a representation of a coordinate system that can be used to reason about the user's surroundings. Each `SpatialCoordinateSystem` has a relationship with other coordinate systems, which means that given the user's current coordinate system, we can work out where the frame is relative to the user's current position. We make use of this to determine where to place the label, which is the work of this section.

Jump back into the `ProcessFrame` method of the `FaceTagMain` class and make the following amendments (shown in bold):

```
void ProcessFrame(SpatialCoordinateSystem worldCoordinateSystem)
{
if (!IsInValidateStateToProcessFrame())
{
return;
}

// obtain the details of the last frame captured
FrameGrabber.Frame frame = frameGrabber.LastFrame;

if (frame.mediaFrameReference == null)
{
return;
}

MediaFrameReference mediaFrameReference = frame.mediaFrameReference;

SpatialCoordinateSystem cameraCoordinateSystem =
mediaFrameReference.CoordinateSystem;
CameraIntrinsics cameraIntrinsics =
mediaFrameReference.VideoMediaFrame.CameraIntrinsics;
```

```
Matrix4x4? cameraToWorld =
cameraCoordinateSystem.TryGetTransformTo(worldCoordinateSystem);

if (!cameraToWorld.HasValue)
{
return;
}

frameAnalyzer.AnalyzeFrame(frame.mediaFrameReference, (status,
detectedPersons) =>
{
if (status > 0 && detectedPersons.Count > 0)
{
FrameAnalyzer.Bounds? bestRect = null;
Vector3 bestRectPositionInCameraSpace = Vector3.Zero;
float bestDotProduct = -1.0f;
FrameAnalyzer.DetectedPerson bestPerson = null;

foreach (var dp in detectedPersons)
{
Debug.WriteLine($"Detected person: {dp.ToString()}");
}
}
});
}
```

In the preceding code snippet, we are getting reference to SpatialCoordinateSystem and CameraIntrinsics of the MediaFrameReference; we then try to transform the frame's coordinates into the user's current coordinate space, done in the statement:

```
Matrix4x4? cameraToWorld =
cameraCoordinateSystem.TryGetTransformTo(worldCoordinateSystem);
```

As mentioned earlier, each coordinate system has a dynamic relationship with the other coordinate systems; this relationship can be encoded as a 3D transformation matrix, returned by the preceding statement if successful. With this matrix, we can transform points and directions from one space (frames coordinate system) into another (user's coordinate system).

Along with the transformation, we add some variables that we will use over the next few paragraphs to find the most relevant detected face and details of where they are located in the world space.

Next, we will define some variables that we will use to position the label; add the following lines to the code you added earlier (before passing the frame to the `FrameAnalzyer`):

```
float averageFaceWidthInMeters = 0.15f;

float pixelsPerMeterAlongX = cameraIntrinsics.FocalLength.X;
float averagePixelsForFaceAt1Meter = pixelsPerMeterAlongX *
averageFaceWidthInMeters;

Vector3 labelOffsetInWorldSpace = new Vector3(0.0f, 0.25f, 0.0f);
```

We first calculate the average number of pixels for a face 1 meter away; we will use this soon to infer the depth of any person we identify. We then define a vector that will be used to offset the label from the center of the detected face; here, we are positioning the label 0.25 meters directly above the face.

Next, we will amend the body of the detected face loop; we restrict this example to display a single person at a time. To select the most relevant person, we calculate the dot product for each face, keeping reference to the largest value (which gives us the detected face that's most aligned with the user's current gazing direction); we step through the following amendments:

```
FrameAnalyzer.Bounds? bestRect = null;
Vector3 bestRectPositionInCameraSpace = Vector3.Zero;
float bestDotProduct = -1.0f;
FrameAnalyzer.DetectedPerson bestPerson = null;
```

We start by reconstructing the ray of the camera to the face when the frame is captured; we will use this to determine the difference in angle (dot product) with respect to the user's current gaze, and thus determine how relevant this person is. Add the following code to create a point to represent the center of the current face detected:

```
Point faceRectCenterPoint = new Point(
dp.bounds.left + dp.bounds.width /2,
dp.bounds.top + dp.bounds.height / 2
);
```

Next, we unproject this from pixel coordinates into a camera space ray from the camera origin, expressed as a x and y coordinates on the plane at $z = 1.0$, and then align the z axis with the user's gaze:

```
Vector2 centerOfFace =
cameraIntrinsics.UnprojectAtUnitDepth(faceRectCenterPoint);
Vector3 vectorTowardsFace = Vector3.Normalize(new Vector3(centerOfFace.X,
centerOfFace.Y, -1.0f));
```

Next, we calculate the dot product that will be used to determine how relevant this face is compared to any other detected faces:

```
float dotFaceWithGaze = Vector3.Dot(vectorTowardsFace, -Vector3.UnitZ);
```

This gives us our dot product between the user's gaze and detected face; we compare our current best with this and if better, update the variables. Check the following code snippet for this:

```
if (dotFaceWithGaze > bestDotProduct)
{
float estimatedFaceDepth = averagePixelsForFaceAt1Meter /
(float)dp.bounds.width;
Vector3 targetPositionInCameraSpace = vectorTowardsFace *
estimatedFaceDepth;

bestDotProduct = dotFaceWithGaze;
bestRect = dp.bounds;
bestRectPositionInCameraSpace = targetPositionInCameraSpace;
bestPerson = dp;
}
```

Along with updating the local variables, we also calculate a position shown in the following lines:

```
float estimatedFaceDepth = averagePixelsForFaceAt1Meter /
(float)dp.bounds.width;
Vector3 targetPositionInCameraSpace = vectorTowardsFace *
estimatedFaceDepth;
```

Once we have iterated through all detected faces, we test whether a face was detected, and if so, update our `QuadRenderer` and `TextRenderer`, also known as our label, as follows:

```
if (bestRect.HasValue)
{
Vector3 bestRectPositionInWorldspace =
Vector3.Transform(bestRectPositionInCameraSpace, cameraToWorld.Value);
Vector3 labelPosition = bestRectPositionInWorldspace +
labelOffsetInWorldSpace;

quadRenderer.TargetPosition = labelPosition;
textRenderer.RenderTextOffscreen($"{bestPerson.name},
{bestPerson.gender}, Age: {bestPerson.age}");

lastFaceDetectedTimestamp = Utils.GetCurrentUnixTimestampMillis();
}
```

Most notable is how we transform the position into the user's space, using the 3D transformation matrix we obtained before between the frames coordinate system and users current coordinate system, which is what we do in the following code snippet:

```
Vector3 bestRectPositionInWorldspace =
Vector3.Transform(bestRectPositionInCameraSpace, cameraToWorld.Value);
Vector3 labelPosition = bestRectPositionInWorldspace +
labelOffsetInWorldSpace;
```

With that, we conclude the example; the full listing of the `ProcessFrame` method is demonstrated here:

```
void ProcessFrame(SpatialCoordinateSystem worldCoordinateSystem)
{
if (!IsInValidateStateToProcessFrame())
{
return;
}

FrameGrabber.Frame frame = frameGrabber.LastFrame;

if (frame.mediaFrameReference == null)
{
return;
}

MediaFrameReference mediaFrameReference = frame.mediaFrameReference;

SpatialCoordinateSystem cameraCoordinateSystem =
mediaFrameReference.CoordinateSystem;
CameraIntrinsics cameraIntrinsics =
mediaFrameReference.VideoMediaFrame.CameraIntrinsics;

Matrix4x4? cameraToWorld =
cameraCoordinateSystem.TryGetTransformTo(worldCoordinateSystem);

if (!cameraToWorld.HasValue)
{
return;
}

float averageFaceWidthInMeters = 0.15f;

float pixelsPerMeterAlongX = cameraIntrinsics.FocalLength.X;
float averagePixelsForFaceAt1Meter = pixelsPerMeterAlongX *
averageFaceWidthInMeters;

Vector3 labelOffsetInWorldSpace = new Vector3(0.0f, 0.25f, 0.0f);
```

```
frameAnalyzer.AnalyzeFrame(frame.mediaFrameReference, (status,
detectedPersons) =>
{
if(status > 0 && detectedPersons.Count > 0)
{
FrameAnalyzer.Bounds? bestRect = null;
Vector3 bestRectPositionInCameraSpace = Vector3.Zero;
float bestDotProduct = -1.0f;
FrameAnalyzer.DetectedPerson bestPerson = null;

foreach (var dp in detectedPersons)
{
Point faceRectCenterPoint = new Point(
dp.bounds.left + dp.bounds.width /2,
dp.bounds.top + dp.bounds.height / 2
);

Vector2 centerOfFace =
cameraIntrinsics.UnprojectAtUnitDepth(faceRectCenterPoint);

Vector3 vectorTowardsFace = Vector3.Normalize(new
Vector3(centerOfFace.X, centerOfFace.Y, -1.0f));

float dotFaceWithGaze = Vector3.Dot(vectorTowardsFace, -
Vector3.UnitZ);

if (dotFaceWithGaze > bestDotProduct)
{
float estimatedFaceDepth = averagePixelsForFaceAt1Meter /
(float)dp.bounds.width;

Vector3 targetPositionInCameraSpace = vectorTowardsFace *
estimatedFaceDepth;

bestDotProduct = dotFaceWithGaze;
bestRect = dp.bounds;
bestRectPositionInCameraSpace = targetPositionInCameraSpace;
bestPerson = dp;
}
}

if (bestRect.HasValue)
{
Vector3 bestRectPositionInWorldspace =
Vector3.Transform(bestRectPositionInCameraSpace, cameraToWorld.Value);
Vector3 labelPosition = bestRectPositionInWorldspace +
labelOffsetInWorldSpace;
```

```
quadRenderer.TargetPosition = labelPosition;
textRenderer.RenderTextOffscreen($"{bestPerson.name},
{bestPerson.gender}, Age: {bestPerson.age}");

lastFaceDetectedTimestamp = Utils.GetCurrentUnixTimestampMillis();
    }
  }
});
}
```

As we did before, turn on your HoloLens device and click on the **Remote Machine** deployment button to deploy and run the application; you should now be able to see metadata floating above the heads of the persons you have registered, as shown in the following image (flattering age prediction from Microsoft Cognitive Services):

Let's conclude this chapter with a quick summary of what we covered before moving on to the next example.

Summary

Congratulations for getting through this chapter. We have covered a lot and only just touched the surface on the topics of DirectX and MR applications, but hopefully enough to excite you. We first walked through the Visual C# DirectX 11 Holographic Template, pointing out some of the main components that are most relevant when developing for the HoloLens. We then built on top of the template, creating an AR application capable of detecting and identifying faces and rendering the details in scene.

As mentioned in the opening of this chapter; I hope this chapter has acted, in part, as a catalyst, sparking your imagination of the possibilities of MR applications. It doesn't take much to extend this example to something that can help the visually impaired better sense and understand their surroundings or help collapse cultural barriers by automatically translating foreign languages into your native tongue.

In the next chapter, we will continue this path of imagining how we can envision new ways we can assist people by building a proof of concept application that will assist the user in finding items by recording what they see and mapping the path when requested.

3
Assistant Item Finder Using DirectX

In the previous chapter, we explored the opportunities that HoloLens opens up by being able to see everything you see; our exploration began by setting up and walking through the holographic DirectX 11 application template Microsoft has bundled in *Visual Studio 2015 Update 3*. We then walked through extending this application to detect and identify faces, and display associated metadata about the identified person the user was looking at.

In this chapter, we continue our exploration with how HoloLens can better assist people through a proof of concept example that continuously observes the world, identifying and tracking salient items that you (it) see. If the user happens to forget where an item is, they will be able to ask for assistance and be guided via a *breadcrumb* trail from the user to the item. For instance, imagine, if you will, a time when you have misplaced your keys and spent vast amounts of time searching for them. Now, imagine a world, in the very near future, where your device can conveniently track your items and lead you to them--possibly adapting the path that best suits your needs, for example, by that is routing you through a wheel chair-friendly path.

Once again, I encourage you to let your imagination wander while you read this chapter, dreaming of how this concept can be extended. One example might involve, identifying what is in the fridge and automatically constructing and presenting a shopping list when the user is in proximity of a grocery store. These are exciting times ahead. With that being said, let's get started!

We will start by introducing the application and all of the major components, then walk through each one until we have a functioning prototype. We begin by mapping out the environment, then tagging the location of the items we detect, and finally, building and rendering the trail from the user to the "lost" item. In this chapter, you will learn the following topics:

- Creating and using `SpatialLocatorAttachedFrameOfReference` to consider the user's location from the perspective of the device
- Making use of **spatial anchors** to map the physical world
- Using **Microsoft Cognitive Services** to recognize objects

Project setup

Our first stop will be looking at how we can make sense of the world, or at least, programmatically be aware of what the user is looking at. Computer vision, specifically recognition, has made leaps and bounds since 2012, when computer scientists Geoffrey Hinton, Alex Krizhevsky, and Ilya Sutskever entered the ILSVRC 2012 computer vision competition using ideas from `http://www.cs.toronto.edu/~fritz/absps/imagenet.pdf`, a paper they had recently published. Being the only ones using a **Convolutional Neural Network** (**CNN**), they entered the competition; the rest is pretty much history. Models these days can compete with humans in recognizing objects in images.

 CNN is a type of neural network well suited for images due to its properties of preserving the relationship between pixels in close proximity.

Fortunately for us, many companies have made **Application Program Interfaces** (**APIs**) available that offer similar capabilities for computer vision, including Microsoft. We will be using Microsoft's Cognitive Services Computer Vision API for this example. The API gives our application the ability to recognize objects in a image, or in our case, in front of the user. We will use this to track what items the user comes across. Before jumping into code, let's see how this API works and what capabilities it provides for us.

The easiest way to do this is to go to their service webpage `https://azure.microsoft.com/ en-gb/services/cognitive-services/computer-vision/` and explore some of the examples presented on this page. The following is one such example showing the image and associated metadata:

Screenshot of the computer vision Microsoft offers as ones of its cognitive services, Source: https://azure.microsoft.com/en-gb/services/cognitive-services/computer-vision

In the preceding image, you can see the service returning a set of **Tags** and **Captions.** This is what we will assign, each frame (captured image) we capture and we will use this data when searching for an item for the user.

Adding the Computer Vision API service

As we did in the previous chapter, to use this service, we will need to add it to our Microsoft Azure portal. Head over to `https://portal.azure.com/` on your browser and log in (or register if you don't already have an account).

Once logged in, you will be taken to your **Dashboard**; from here, click on the **+New** button at the top of the left panel. This will expand out a menu. Within this menu, search and click on **AI + Cognitive Services**, which will spawn another menu where we can select the **Computer Vision API**, as shown in the following screenshot:

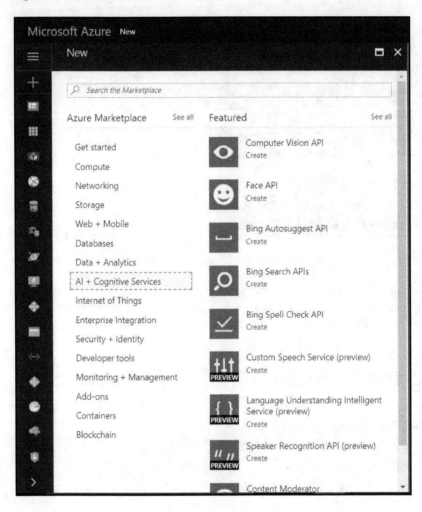

After clicking on the **Computer Vision API** item, you will be presented with a tab requiring you to enter the details of your service, such as **Name**, **Subscription**, **Location**, **Pricing tier**, and **Resource Group**, and help is accessible within the user interface. With all that done, you have your service up and running; details of this service should be visible on the screen and can be accessed from your dashboard, as shown in the following screenshot:

To use your service, you will need to call it using the generated keys. These can be accessed by clicking on **Show access keys...** which, when clicked, will display a set of keys you will need. For convenience, copy and paste them somewhere easily accessible (Notepad) for use later in this chapter.

The project

To recap, the objectives for this application are as follows:

- Plot out the environment so we can reconstruct a path from the user's location to the specified item
- Continuously process frames captured by the HoloLens color camera and attach the tags returned by Microsoft's Cognitive Service Vision API

But first, let's set up the environment and then look at how we can better understand the environment by plotting paths where the user walks using spatial anchors (via `SpatialAnchor`). Start by launching the starter solution for this chapter; this can be found in the directory *Chapter3/Starter* from the repository `https://github.com/PacktPublishing/Microsoft-HoloLens-By-Example`. Launch the Visual Studio solution by either double-clicking on the solution file `AssistantItemFinder.sln` or by loading from within Visual Studio by navigating via **File** | **Open** | **Project** | **Solution** and selecting the `AssistantItemFinder.sln` file. The starter solution is more or less the standard Visual C# holographic DirectX 11 app template that we used in the previous chapter, with the addition of some classes and resources used to render the trail for the user to follow.

Before we start coding, let's get some administration out of the way. The first is to set the appropriate **Capabilities** for this project, which is access to the color camera (**Webcam**) and Internet (**Internet (Client)**). Open up the **Capabilities** panel of the package properties by double-clicking on the `Package.appxmanifest` file and selecting the **Capabilities** tab. Once open, check the capabilities:

- **Internet (Client)**
- **Webcam**

Your **Capabilities** panel should look similar to the screenshot with the previously mentioned capabilities checked:

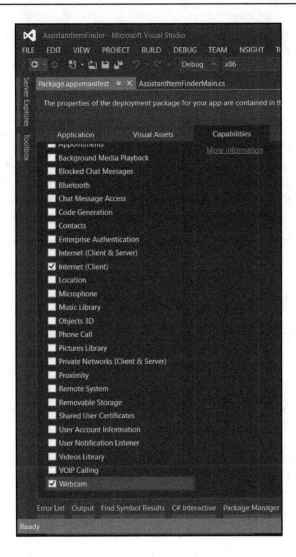

Next, we will amend the `cognitive_service.key` file that resides in your projects directory with the key from the service we just created. If you have not done so already, copy the key from the service we created earlier and paste it into this file. We use this when we build the class responsible for calling the Computer Vision API.

With the setup now out of the way, we can move onto building our application. In the next section, we will add the functionality to build a network of nodes and edges that we will use to construct walkable paths for the user when navigating them to an item.

Leaving breadcrumbs

The solution to this example was inspired in part by the fairy tale *Hansel and Gretel*, specifically, the part where the main characters in the story leave a trail of breadcrumbs to keep track of their location as they are led deep into the forest to be left abandoned. We will implement a similar strategy here; as the user walks, we will drop nodes to keep track of their path, as illustrated in the following figure:

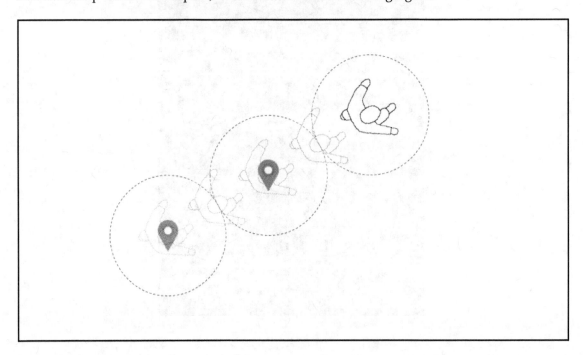

Let's begin our journey--start by updating `referenceFrame` from
`SpatialStationaryFrameOfReference` to
`SpatialLocatorAttachedFrameOfReference`. Open up the
file `AssistantItemFinderMain.cs` and navigate down to where `referenceFrame` is
declared and change its class from `SpatialStationaryFrameOfReference` to
`SpatialLocatorAttachedFrameOfReference`, as follows:

```
    internal class AssistantItemFinderMain : IDisposable,
IFrameGrabberDataSource
    {
        ...
        SpatialLocatorAttachedFrameOfReference referenceFrame;
        ...
    }
```

Now, make the following changes to the `SetHolographicSpace` method:

```
public void SetHolographicSpace(HolographicSpace holographicSpace)
{
  this.holographicSpace = holographicSpace;
  locator = SpatialLocator.GetDefault();
  locator.LocatabilityChanged += this.OnLocatabilityChanged;
  holographicSpace.CameraAdded += this.OnCameraAdded;
  holographicSpace.CameraRemoved += this.OnCameraRemoved;
  referenceFrame=
  locator.CreateAttachedFrameOfReferenceAtCurrentHeading();
}
```

As you may remember from the previous chapter, `referenceFrame` is used by the
HoloLens to consider the position and the orientation of the holograms it is rendering. In
the previous chapter, we used a stationary reference frame that
is `SpatialStationaryFrameOfReference`--which, once created, was static (until
recreated). All the holograms would be positioned relative to this specific position. In this
chapter, we are using an attached reference frame that
is `SpatialLocatorAttachedFrameOfReference`--where, as the name implies, this origin
is the position of the HoloLens device that is updated as the device moves.

The position of `SpatialLocatorAttachedFrameOfReference` is
updated to match the position of the HoloLens, but the orientation
remains static.

Changing the type of `referenceFrame` will introduce many compile time errors. Let's jump through the code, making the necessary changes to resolve these.

First jump to the `Update` method and change the following statement:

```
SpatialCoordinateSystem currentCoordinateSystem =
referenceFrame.CoordinateSystem;
```

This will make it look as follows:

```
SpatialCoordinateSystem referenceFrameCoordinateSystem =
referenceFrame.GetStationaryCoordinateSystemAtTimestamp(prediction.Timestam
p);
```

Similarly, with the `Render` method, make the following changes from the following line:

```
cameraResources.UpdateViewProjectionBuffer(deviceResources, cameraPose,
referenceFrame.CoordinateSystem);
```

Change the preceding line to the following code snippet:

```
SpatialCoordinateSystem referenceFrameCoordinateSystem =
referenceFrame.GetStationaryCoordinateSystemAtTimestamp(prediction.Timestam
p);

if(referenceFrameCoordinateSystem == null)
{
 continue;
}

cameraResources.UpdateViewProjectionBuffer(deviceResources, cameraPose,
referenceFrameCoordinateSystem);
```

With the reference frame now updated, it's time to create our breadcrumbs--as discussed in the previous chapter, the HoloLens works over large areas. Unlike VR that can use a single coordinate system, the HoloLens needs to spawn these across a physical space. This makes sense, given that the HoloLens uses the physical world for reference; that is, if the HoloLens travels too far from a recognizable point, then it will be unable to accurately consider its position. For this reason, the `Windows.Perception.Spatial` namespace includes something called `SpatialAnchor`. `SpatialAnchor` allows us to explicitly create reference points, and therefore, allows us to extend it beyond its initial area. When `SpatialAnchor` is created, relationships are created between itself and other coordinate systems, allowing us to transform between them--something you will become familiar with by the end of this chapter. You may be wondering--*how is this related to our breadcrumbs?*--our breadcrumbs will essentially be `SpatialAnchor` with some additional metadata.

We will soon jump back in the editor and create these breadcrumbs (here on, referred to as nodes), but before we do, let's first discuss the logic behind creating them.

Our requirement is to be able to rebuild the user's path back to an item. To avoid having to scan the physical space and perform some type of path finding algorithm over the physical space, we will simply build a dense mesh as the user traverses the environment. Each node will reference an associated `SpatialAnchor` along with a collection of edges that will connect nearby nodes the user travels to and from. When constructing a path for the user, we can assume that these edges provide us with a clear path. The following figure illustrates this concept:

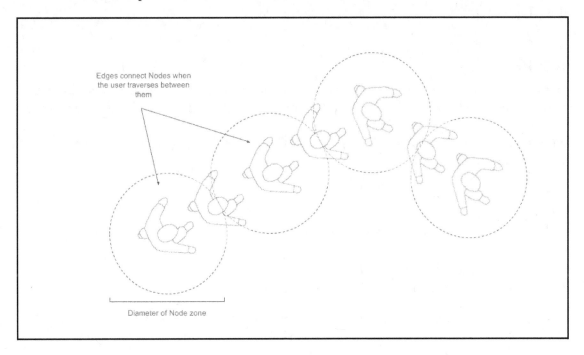

As illustrated in the preceding figure, nodes are created when the user's distance is beyond a predefined distance from the current node that the user resides in. When this occurs, the application will search for a node in proximity, and create a new node only if one cannot be found. When the user traverses between nodes, an edge is created that will connect them and, thus, make a path available for us to use.

During this brief discussion, we have identified a set of entities we will need to create. The following is a list of each of them with a short explanation of their purpose:

- **Node**: This is the breadcrumb of our system--these will be dropped as the user walks around their environment
- **Edge**: This object creates a connection between nodes
- **Entity**: This is the visual representation of a node and is used when we, obviously, want to render a path

Let's begin by creating our `Node` and `Edge` classes and then implement the logic to manage this process in the `AssistantItemFinderMain` class.

Node

Within Visual Studio, create a new class called `Node` in the `Content` folder. This class acts as our breadcrumb, which is essentially a wrapper around `SpatialAnchor`. To gain access to `SpatialAnchor`, we must import the following namespace: `Windows.Perception.Spatial`. Copy the following snippet to the newly created `Node` class:

```
internal class Node
{
   static int NodeID = 0;
   public string Name { get; private set; }
   public Vector3 Position { get; set; }
   public Vector3 Forward { get; set; }
   public long Timestamp { get; set; }
   public SpatialAnchor Anchor { get; set; }

   public Node(SpatialAnchor anchor, Vector3 position, Vector3 forward)
     {
       Name = $"Node_{++Node.NodeID}";
       Anchor = anchor;
       Position = position;
       Forward = forward;
       Timestamp = Utils.GetCurrentUnixTimestampMillis();
     }
}
```

In the preceding code snippet, we have created the variables that Node will need to track, the main one being the SpatialAnchor. The properties Position and Forward are the position and gaze direction of the user relative to the assigned SpatialAnchor. Along with the variables, we will also add some helper methods. Let's add each, one by one, starting with GetTransform, as shown in the following snippet:

```
public Matrix4x4? GetTransform(SpatialCoordinateSystem
targetCoordinateSystem)
{
var mat =
Anchor.CoordinateSystem.TryGetTransformTo(targetCoordinateSystem);
if (!mat.HasValue)
{
  return null;
}
Matrix4x4 modelTranslation = Matrix4x4.CreateTranslation(Position);
return (modelTranslation) * mat.Value;
}
```

Our first method, GetTransform, will return the transformation matrix from the assigned spatial anchor's coordinate system to targetCoordinateSystem. This can then be used to transform holograms that are anchored to the assigned SpatialAnchor to the coordinate system passed in targetCoordinateSystem. You will see how we use this when we come to render each Node:

```
public Vector3? TryGetTransformedPosition(SpatialCoordinateSystem
targetCoordinateSystem)
{
  if (targetCoordinateSystem == Anchor.CoordinateSystem)
  {
   return Position;
  }
var trans =
Anchor.CoordinateSystem.TryGetTransformTo(targetCoordinateSystem);
if (trans.HasValue)
  {
   return Vector3.Transform(Position, trans.Value);
  }
return null;
}
```

The method `TryGetTransformedPosition` is very similar to the previous method, `GetTransform`; however instead of returning a transformation matrix, it will return `Position` relative to `targetCoordinateSystem` passed in as a parameter. This method will be used when determining the position of `Node` relative to the reference frame's position (the reference frame being the spatial details associated to the HoloLens device). As we did previously, we obtain the transformation matrix via the line `var trans = Anchor.CoordinateSystem.TryGetTransformTo(targetCoordinateSystem);`; now, armed with the transformation matrix, we simply transform the position of `Node` to `targetCoordinateSystem` using the `Vector3.Transform` method, as follows:

```
public float? TryGetDistance(SpatialCoordinateSystem
targetCoordinateSystem, Vector3 targetPosition)
{
  if (targetCoordinateSystem == Anchor.CoordinateSystem)
  {
   return (targetPosition - Position).Length();
  }

  var trans =
  Anchor.CoordinateSystem.TryGetTransformTo(targetCoordinateSystem);
  if (trans.HasValue)
  {
   return (targetPosition - Vector3.Transform(Position,
   trans.Value)).Length();
  }
  return null;
}
```

The `TryGetDistance` method returns the euclidean distance between two positions relative to `targetCoordinateSystem`. The method starts by checking whether the coordinate system of `Node` is the same as the coordinate system passed in and, if so, returns the distance between the two positions with no further calculation. Otherwise, it attempts to create a transformation matrix from the anchor's coordinate system and the coordinate system passed in. If successful, it will transform the position of `Node` before calculating and returning the distance.

With that method now complete, this concludes our `Node` class, for now at least. Next, we will turn our attention to the `Edge` class.

Edges

As we did before, create a new class called `Edge` in the `Content` folder of the `AssistantItemFinder` project. Within the new class open, add the namespace `System.Numerics` to get access to some of the extended Math types, such as `Vector3`, and make the following amendments:

```
internal class Edge
{
public Node NodeA { get; set; }
public Node NodeB { get; set; }

public Vector3 Direction
{
get{
if (NodeA.Anchor == NodeB.Anchor)
{
  return Vector3.Normalize(NodeB.Position - NodeA.Position);
}
else{
var nodeBToBTrans =
NodeB.Anchor.CoordinateSystem.TryGetTransformTo(NodeA.Anchor.Coordinat
eSystem);
if (nodeBToBTrans.HasValue)
{
var nodeBPosition = Vector3.Transform(NodeB.Position,
nodeBToBTrans.Value);
return Vector3.Normalize(nodeBPosition - NodeA.Position);
}
}
return Vector3.Zero;
}
}

public float Distance
{
get{
if (NodeA.Anchor == NodeB.Anchor)
{
return (NodeB.Position - NodeA.Position).Length();
}
else{
var nodeBToBTrans =
NodeB.Anchor.CoordinateSystem.TryGetTransformTo(NodeA.Anchor.Coordinat
eSystem);
if (nodeBToBTrans.HasValue)
{
```

```
var nodeBPosition = Vector3.Transform(NodeB.Position,
nodeBToBTrans.Value);
return (nodeBPosition - NodeA.Position).Length();
}
}
return 0f;
}
}


}
```

As mentioned previously, the `Edge` class is used to connect two nodes when the user transverses between them. We capture this connection by referencing the source node (`NodeA`) and the destination node (`NodeB`). In addition, we add two helper properties that return `Angle` and `Distance` between the two nodes.

The next class we implement can be considered to be the visual representation of `Node`. Its sole purpose is to render `Node` using the assigned `Renderer`. After implementing this, we will return to the `AssistantItemFinderMain` class to hook everything together.

Entity

The purpose of the `Entity` class is to hold reference `Renderer` and `Node`, and use the position of `Node` and the orientation to update the buffer of `Renderer` before rendering.

Create a new class called `Entity` in the `Content` folder of the `AssistantItemFinder` project. With the new class open, add the namespace `System.Numerics`, `Windows.Perception.Spatial`, and `AssistantItemFinder.Common` and make the following amendments to the class:

```
internal class Entity
{
  public string Tag { get; private set; }
  public Node Node { get; set; }
  public Renderer Renderer { get; set; }

  public Vector3 Position { get; set; }
  public Vector3 EulerAngles { get; set; }
  public Vector3 Scale { get; set; }

  public bool Visible { get; set; }
  public bool Enabled { get; set; }

  private Matrix4x4 transform = Matrix4x4.Identity;
  public Matrix4x4 Transform
```

```
{
get{
return transform;
}
}

public Entity(string tag, bool visible = true, bool enabled = true)
{
Tag = tag;
Visible = visible;
Enabled = enabled;

Position = Vector3.Zero;
EulerAngles = Vector3.Zero;
Scale = Vector3.One;
}
}
```

In the preceding snippet, we have created the properties and constructor for our
Entity class; no surprises here. Entity is responsible for rendering Node using the
assigned Renderer. Entity itself contains Position and Rotation, all of which are
relative to the anchor's coordinate system of Node and applied to the transformation when
rendering. Next, we will add the Update method responsible for updating the instance's
transform variable using the one passed in the coordinate system. Add the following
method to your Entity class:

```
public virtual void Update(StepTimer timer, SpatialCoordinateSystem
referenceFrameCoordinateSystem)
{
if(!Enabled){ return; }

UpdateTransform(referenceFrameCoordinateSystem);
}

public void UpdateTransform(SpatialCoordinateSystem
referenceFrameCoordinateSystem)
{
 var trans = Node.GetTransform(referenceFrameCoordinateSystem);
 if (trans.HasValue)
 {
  Matrix4x4 modelTranslation = Matrix4x4.CreateTranslation(Position);
  Matrix4x4 modelRotation = Matrix4x4.CreateFromYawPitchRoll(
  DegreeToRadian(EulerAngles.Y),
  DegreeToRadian(EulerAngles.X),
  DegreeToRadian(EulerAngles.Z));
  Matrix4x4 modelScale = Matrix4x4.CreateScale(Scale);
```

```
    transform = (modelScale * modelRotation * modelTranslation) *
    trans.Value;
  }
}

private float RadianToDegree(double angle)
{
return (float)(angle * (180.0 / Math.PI));
}
```

The `UpdateTransform` method is the main workhorse in the preceding snippet. It's responsible for constructing the model matrix `transform` using the `Entity`, `Scale`, `Rotation`, and `Position` properties, as well as taking into account the transformation matrix between the referenced coordinate system of `Node` via `SpatialAnchor` and the coordinate system passed in which will be the coordinate system associated to the reference frame.

The following snippet contains the final piece of code for our `Entity` class and is responsible for updating the assigned model matrix of `Renderer` before delegating the rendering to it:

```
public virtual void Render()
{
  if (!Visible || Renderer == null){ return; }

  Renderer.UpdateModelTransform(Transform);
  Renderer.Render();
}
```

With the `Node`, `Edge`, and `Entity` classes now implemented, we will return to the `AssistantItemFinderMain` class and hook everything up.

Leaving a trail

Start by adding the following namespaces to the `AssistantItemFinderMain` class: `System.Numerics`, `Windows.Perception`, and `AssistantItemFinder.Content` as well as adding the following variables and properties:

```
const float EntityOffsetY = -1.0f;
const float NodeRadius = 0.8f;

private Node currentNode = null;
private Vector3 currentNodePosition = Vector3.Zero;
private Vector3 currentGazeForward = Vector3.Zero;
private Vector3 currentGazeRight = Vector3.Zero;
```

```
private double dwellTimeAtCurrentNode = 0f;

private List<Node> nodes = new List<Node>();
private List<Edge> edges = new List<Edge>();
private List<Entity> entities = new List<Entity>();

private Renderer nodeRenderer;

public Node CurrentNode
{
get{
lock (this)
{
return currentNode;
}
}
set{
lock (this)
{
if(currentNode != value)
{
dwellTimeAtCurrentNode = 0;
currentNode = value;
}
}
}
}

public Vector3 NodePosition
{
get{
lock(this)
{
return currentNodePosition;
}
}
set{
lock(this)
{
currentNodePosition = value;
}
}
}

public Vector3 GazeForward
{
get{
lock(this)
```

```
{
return GazeForward;
}
}
set{
lock(this)
{
GazeForward = value;
}
}
}

public Vector3 GazeUp
{
get{
lock(this)
{
return currentGazeRight;
}
}
set{
lock(this)
{
currentGazeRight = value;
}
}
}
```

Most should be self-explanatory and following are some details for those that require some explanation:

- **EntityOffsetY**: It is an offset applied to the y axis of the entity when rendering. This offset controls the height that the nodes are rendered.
- **NodeRadius**: It specifies how large we want the zone of a node to be; reducing it will increase the density of our graph, while increasing it will make it more sparse and less accurate--obviously. This number was selected arbitrarily but can be easily tuned.
- **dwellTimeAtCurrentNode**: It is used to store how long the user has resided in a single node. This is used later on in this chapter when we start dealing with mapping the path for the user.

Next, we will deal with the creation and destruction of `Renderer`. Now, within the `SetHolographicSpace` method, make the following amendments:

```
nodeRenderer = new NodeRenderer(deviceResources, new Vector3(0.3f, 0.3f,
1.0f), 0.05f);
nodeRenderer.CreateDeviceDependentResourcesAsync();
```

And being good citizens, we will destroy it in the right places. Make the following amendments to the `Dispose` method:

```
public void Dispose()
{
...

if (nodeRenderer != null)
{
nodeRenderer.Dispose();
nodeRenderer = null;
}
}
```

With the variables and properties out of the way, it's time to hook everything together. We start in the `Update` method where everything is driven from. After this, we will build out the methods we need to support our task--developing in a type of top-down approach.

Within the `Update` method, under the state, make the following amendments from the snippets that follow:

```
var previousNode = CurrentNode;
CurrentNode = UpdateCurrentNode(referenceFrameCoordinateSystem,
prediction.Timestamp, NodeRadius);
```

We first call `UpdateCurrentNode`, which will be responsible for finding the closest node, otherwise, creating a new one before returning it:

```
SpatialPointerPose pose =
SpatialPointerPose.TryGetAtTimestamp(referenceFrameCoordinateSystem,
prediction.Timestamp);

NodePosition = pose.Head.Position;
GazeForward = pose.Head.ForwardDirection;
GazeUp = pose.Head.UpDirection;

var mat =
referenceFrameCoordinateSystem.TryGetTransformTo(CurrentNode.Anchor.Coordin
ateSystem);
if (mat.HasValue)
```

```
{
NodePosition = Vector3.Transform(NodePosition, mat.Value);
GazeForward = Vector3.TransformNormal(GazeForward, mat.Value);
GazeUp = Vector3.TransformNormal(GazeUp, mat.Value);
}
```

In the preceding snippet, we are updating our properties associated to the user's current position and pose. These are made available via the `SpatialPointerPose` class obtained by the
statement `SpatialPointerPose.TryGetAtTimestamp(referenceFrameCoordinateSystem, prediction.Timestamp)` and is relative to the coordinate system and predicted time we pass in as arguments. We then try to obtain the transformation matrix between the reference frame's coordinate system and the current node's coordinate system. If successful, use it to transform the position and pose associated with the reference frame's coordinate system into the coordinate system of the current node:

```
CreateEntitiesForAllNodes();

timer.Tick(() =>
{
dwellTimeAtCurrentNode += timer.ElapsedSeconds;

for (var entityIndex=0; entityIndex<entities.Count; entityIndex++)
{
var entity = entities[entityIndex];
entity.Update(timer, referenceFrameCoordinateSystem);
}
}
```

In the preceding snippet, we make use of a temporary method `CreateEntitiesForAllNodes` that, as you may have guessed, creates an entity for each node. This is so we can test that the trails are working correctly. Then, for each entity, we call its `Update` method, as described in the preceding code snippet.

Your `Update` method should now look similar to the following code snippet:

```
public HolographicFrame Update()
{
HolographicFrame holographicFrame = holographicSpace.CreateNextFrame();
HolographicFramePrediction prediction = holographicFrame.CurrentPrediction;
deviceResources.EnsureCameraResources(holographicFrame, prediction);
SpatialCoordinateSystem referenceFrameCoordinateSystem =
referenceFrame.GetStationaryCoordinateSystemAtTimestamp(prediction.Timestamp);

var previousNode = CurrentNode;
```

```
CurrentNode = UpdateCurrentNode(referenceFrameCoordinateSystem,
prediction.Timestamp, NodeRadius);

SpatialPointerPose pose =
SpatialPointerPose.TryGetAtTimestamp(referenceFrameCoordinateSystem,
prediction.Timestamp);

NodePosition = pose.Head.Position;
GazeForward = pose.Head.ForwardDirection;
GazeUp = pose.Head.UpDirection;

var mat =
referenceFrameCoordinateSystem.TryGetTransformTo(CurrentNode.Anchor.Coordin
ateSystem);
if (mat.HasValue)
{
NodePosition = Vector3.Transform(NodePosition, mat.Value);
GazeForward = Vector3.TransformNormal(GazeForward, mat.Value);
GazeUp = Vector3.TransformNormal(GazeUp, mat.Value);
}

CreateEntitiesForAllNodes(referenceFrameCoordinateSystem);

timer.Tick(() =>
{
dwellTimeAtCurrentNode += timer.ElapsedSeconds;

for (var entityIndex=0; entityIndex<entities.Count; entityIndex++)
{
var entity = entities[entityIndex];
entity.Update(timer, referenceFrameCoordinateSystem);
}
}

foreach (var cameraPose in prediction.CameraPoses)
{
HolographicCameraRenderingParameters renderingParameters =
holographicFrame.GetRenderingParameters(cameraPose);
}

return holographicFrame;
}
```

With our Update method now complete, let's move onto the UpdateCurrentNode method; this method is responsible for returning the most appropriate node based on its proximity to the user and, if one doesn't exist, creating it.

It starts by first obtaining the user's current position using the method `SpatialPointerPose.TryGetAtTimestamp` and then testing if `CurrentNode` exists. If this is null, then it is assumed that this is the first call to the method. Therefore, it creates a node. Let's do this now. Add the following code snippet to your `AssistantItemFinderMain` class:

```
Node UpdateCurrentNode(SpatialCoordinateSystem
referenceFrameCoordinateSystem, PerceptionTimestamp perceptionTimestamp,
float nodeRadius = 1.0f)
{
SpatialPointerPose pose =
SpatialPointerPose.TryGetAtTimestamp(referenceFrameCoordinateSystem,
perceptionTimestamp);

if(pose == null){ return; }

if (currentNode == null){
var nodeAnchor = Spatial
Anchor.TryCreateRelativeTo(referenceFrameCoordinateSystem,
pose.Head.ForwardDirection * 0.1f);
if (nodeAnchor == null)
{
returun null;
}
AddNode(nodeAnchor, perceptionTimestamp);
return nodes[nodes.Count-1];
}
}
```

Our `UpdateCurrentNode` method takes the reference frame's coordinate system, the predicted timestamp, and the radius of a node that is the area this node occupies for its arguments. We next obtain the user's current position and pose and create a new node if `CurrentNode` is null. When creating a new node, we first ask the `SpatialAnchor` class to create a new `SpatialAnchor` relative to the reference frame (for the attached frame of reference, this is the currently predicted position of the HoloLens). If successful, we delegate the task to the method `AddNode` passing in the newly created `SpatialAnchor` and predicted timestamp before returning the last node.

One reason that creating `SpatialAnchor` can fail, is if the device has an insufficient understanding of the environment, remembering that it is a physical anchor into the real world and, therefore, requiring enough of the surroundings to have been scanned before being able to construct a unique and robust signature for the anchor.

If the `CurrentNode` is not null, then test how far the center of the node is from the user's position--if within the specified radius, then we return `CurrentNode` with no additional work. Make these changes now by appending the following snippet to your `UpdateCurrentNode` method:

```
else{
var distance = currentNode.TryGetDistance(referenceFrameCoordinateSystem,
pose.Head.Position);
if(distance.HasValue && distance.Value > nodeRadius)
{
 // NEXT SNIPPET GOES HERE
}
return currentNode;
}
```

If the user is out of bounds of `CurrentNode`, we attempt to find a node that satisfies our criteria that is in proximity based on the `nodeRadius` argument passed in. Make the following amendments to the `UpdateCurrentNode` method, continuing from earlier:

```
var closestNodes = GetClosestNodes(referenceFrameCoordinateSystem, pose,
nodeRadius);
if(closestNodes != null && closestNodes.Count > 0)
{
foreach(var node in closestNodes)
{
if(node == currentNode)
{
continue;
}
return node;
}
}
```

The majority of the work is delegated to the method `GetClosestNodes`, to which we will return soon. Failing to find a suitable node, our final option is to create a new node. This is similar to the preceding snippet, where a node was created if none existed with the difference being that we project the node's centroid forward in the direction of the user's gaze and connect `CurrentNode` to this node with an edge.

The following figure illustrates this process:

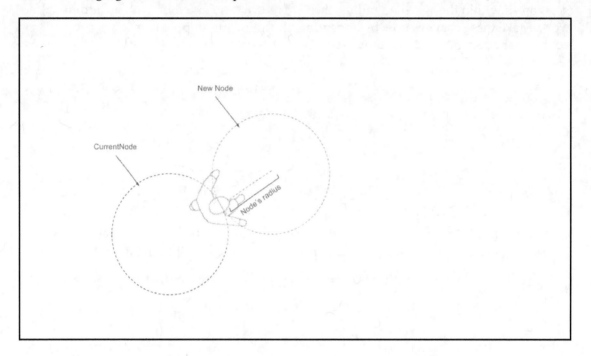

Amend `UpdateCurrentNode` with the following code, starting at the end of the block of the preceding snippet:

```
var currentNodesPosition =
currentNode.TryGetTransformedPosition(referenceFrameCoordinateSystem);
if (currentNodesPosition.HasValue)
{
var direction = Vector3.Normalize(new Vector3(pose.Head.Position.X, 0f,
pose.Head.Position.Z) - new Vector3(currentNodesPosition.Value.X, 0f,
currentNodesPosition.Value.Z));

var targetPosition = currentNodesPosition.Value + direction * nodeRadius;
var distanceFromPose = (targetPosition - new Vector3(pose.Head.Position.X,
0f, pose.Head.Position.Z)).Length();

var nodeAnchor = Spatial
Anchor.TryCreateRelativeTo(referenceFrameCoordinateSystem, (direction *
distanceFromPose));

if (nodeAnchor != null)
{
```

```
var newNode = AddNode(nodeAnchor, perceptionTimestamp);

edges.Add(new Edge{NodeA = currentNode, NodeB = newNode});

return nodes[nodes.Count - 1];
    }
}
```

As mentioned earlier, the major differences between this, and when we previously created a node, have to do with where we position `SpatialAnchor` along with the additional step of connecting `CurrentNode` with the newly created node using an edge.

To further highlight the differences, the following is the snippet related to the positioning of `SpatialAnchor`:

```
var direction = Vector3.Normalize(new Vector3(pose.Head.Position.X, 0f,
pose.Head.Position.Z) - new Vector3(currentNodesPosition.Value.X, 0f,
currentNodesPosition.Value.Z));
var targetPosition = currentNodesPosition.Value + direction * nodeRadius;
var distanceFromPose = (targetPosition - new Vector3(pose.Head.Position.X,
0f, pose.Head.Position.Z)).Length();

var nodeAnchor = Spatial
Anchor.TryCreateRelativeTo(referenceFrameCoordinateSystem, (direction *
distanceFromPose));
```

We first project out in front of the user to determine their facing direction, independent of the *y* axis. We then calculate the target position based on the current node's center position, inferred direction, and specified radius for a node. Finally, we modify the distance, taking into account the current position of the user, rather than the alternative of using the absolute radius.

Creating an edge is simply a matter of instantiating an edge passing in the method `CurrentNode` and the newly created node (`newNode`), as shown in the following code snippet:

```
edges.Add(new Edge{NodeA = currentNode, NodeB = newNode});
```

As a reminder, edges are what we use to construct the path the user will walk to get to their destination. This now completes our `UpdateCurrentNode` method (maybe, a more appropriate name would be `GetNextNode`). Let's now implement the dependencies, which includes the methods `AddNode` and `GetClosestNodes`.

The `AddNode` method takes in the newly created spatial anchor and the perception timestamp. We then obtain the position and pose as we did before, but this time, using the anchor's coordinate system rather than the frame of reference's coordinate system. With these variables, we proceed to create a new node, add it to the nodes collection, and finally, return it to the caller. Add the following method to your `AssistantItemFinderMain` class:

```
Node AddNode(SpatialAnchor anchor, PerceptionTimestamp perceptionTimestamp)
{
var position = Vector3.Zero;
var forward = Vector3.Zero;

var anchorPose =
SpatialPointerPose.TryGetAtTimestamp(anchor.CoordinateSystem,
perceptionTimestamp);

if (anchorPose != null)
{
position = anchorPose.Head.Position;
forward = anchorPose.Head.ForwardDirection;
}

var node = new Node(anchor, position, forward);
nodes.Add(node);

return node;
}
```

Let's now implement the last dependencies for `UpdateCurrentNode`, which will be the `GetClosestNodes` method. This method, as the name suggests, returns a list of nodes ordered by their distance from the user. Add the following method to the `AssistantItemFinderMain` class:

```
public IList<Node> GetClosestNodes(SpatialCoordinateSystem
referenceFrameCoordinateSystem, SpatialPointerPose pose, float nodeRadius =
0.5f)
{
return nodes.OrderBy(node =>
{
return node.TryGetDistance(referenceFrameCoordinateSystem,
pose.Head.Position);
}).Where(node =>
{
return node.TryGetDistance(referenceFrameCoordinateSystem,
pose.Head.Position) <= nodeRadius;
}).ToList();
}
```

We use the convenience of LINQ to order and filter our list. For each node, we are transforming its position into the reference frame's coordinate system and then obtaining its distance from the user. We then filter out nodes that are outside the specified radius distance.

 Language Integrated Query (LINQ) is a Microsoft .NET Framework component that adds query expressions, which are akin to SQL statements, and can be used to conveniently extract and process data from arrays, enumerable classes, XML documents, relational databases, and third-party data sources. To learn more about LINQ, visit the official documentation at https://msdn.microsoft.com/en-us/library/bb308959.aspx.

Now, returning to the Update method, its final dependency is the method CreateEntitiesForAllNodes. Let's add this now to the AssistantItemFinder class:

```
void CreateEntitiesForAllNodes(SpatialCoordinateSystem
referenceFrameCoordinateSystem)
{
 entities.Clear();

 var rootEntity = GetRootEntity(referenceFrameCoordinateSystem);
 var i = 0;
 foreach (var node in this.nodes)
 {
  var entity = new Entity($"node_{i}");
  entity.Node = node;
  entity.Renderer = nodeRenderer;
  entity.UpdateTransform(referenceFrameCoordinateSystem);
  var targetPosition = rootEntity.Transform.Translation;
  entity.Position = new Vector3(0, (targetPosition -
  entity.Transform.Translation).Y, 0f);
  entities.Add(entity);

  i += 1;
 }
}
```

As mentioned earlier, the `CreateEntitiesForAllNodes` method is a temporary method that will create entities for all the created nodes, providing valuable feedback when assessing how well our application is mapping the environment. Here, we simply iterate over each node and create an associative entity. There is one notable step that is worth highlighting; we want to render our nodes at the same height (*y* axis), but because we don't have a single coordinate system, we have to revert to alternative techniques. Here, we are using a root entity (`rootEntity`) as our landmark. After creating this, we offset each entity using the difference between its *y* position and the current node's *y* position. Let's now implement the `GetRootEntity` method and add the following to the `AssistantItemFinder` class:

```
Entity GetRootEntity(SpatialCoordinateSystem
referenceFrameCoordinateSystem)
{
  var entity = new Entity($"node_root");
  entity.Node = nodes[0];
  entity.Renderer = nodeRenderer;
  entity.Position = new Vector3(0, EntityOffsetY, 0);
  entity.UpdateTransform(referenceFrameCoordinateSystem);

  return entity;
}
```

Here, we create a dummy entity (a dummy that we don't intend on using) and set its *y* position equal to the `EntityOffsetY` constant we declared previously. Lastly, we ensure the entity's model matrix is updated by explicitly calling the `UpdateTrasnform` method.

With this method added, we now have an application that will leave a trail where the user walks--let's test it out. Click on **Local Machine** to update the target device. From the dropdown, select the **Remote Machine** option, which will display the following dialog. Ensure the HoloLens is *on* and set its IP address to the value of the **Address** field. Click again on **Remote Machine** to initiate the build and deployment of your application to the HoloLens.

Once deployed and running, walk around your environment to create a trail:

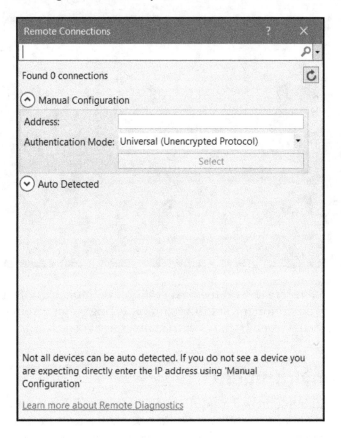

 If this is the first time you are deploying to the HoloLens, it'll ask that you pair the devices. This is just a matter of entering the PIN displayed on the device in the dialog on your computer.

The following figure shows an example of the trail:

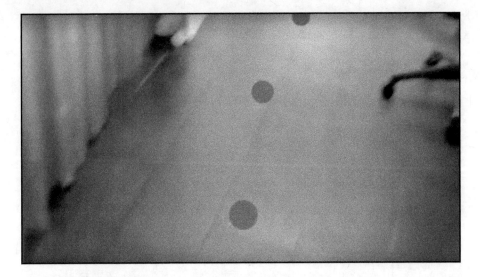

We now have a trail being created. In the next section, we implement the functionality to better understand the environment, starting with capturing frames from the HoloLens camera and then analyzing each frame to extract the salient objects.

Seeing the environment

In this section, we will implement a class that will be responsible for capturing frames from the HoloLens color camera and another class that will make use of Microsoft's Cognitive Vision API service to tag each of these frames to help us understand what the user sees.

Let's start by creating a data object that will be used to store the details of each frame. This data object, Sighting, will be added to a collection of the Node class so that we can find the most appropriate node and construct a path when required.

Within Visual Studio, create a new class called Sighting in the Content folder and make the following amendments:

```
internal class Sighting
{
public string Description { get; private set; }
public HashSet<string> Tokens = new HashSet<string>();

public Vector3 Position { get; private set; }
public Vector3 Forward { get; private set; }
```

```
public Vector3 Up { get; private set; }

public long Timestamp { get; private set; }

public Sighting(string description, Vector3 position, Vector3 forward,
Vector3 up, params string[] tokens)
{
Description = description;

Position = position;
Forward = forward;
Up = up;

if (tokens != null)
{
foreach (var token in tokens)
{
 Tokens.Add(token);
}
}
Timestamp = Utils.GetCurrentUnixTimestampMillis();
}

public Sighting AddToken(string token)
{
Tokens.Add(token);
return this;
}
}
```

In addition, add the namespace `System.Numerics`. As mentioned earlier, the
`Sighting` class simply stores metadata associated with each frame captured from the
HoloLens color camera. As we saw before, Microsoft's Cognitive Vision API returns a list of
tags and a caption for a given photo. We store these in the `tokens` collection and
`description` along with the position and gaze direction relative to the node's anchors
coordinate system. This allows us to determine the general direction of an item (tag) as well
as avoid unnecessary work by analyzing frames that are relatively in the same position and
direction. Let's now make some amendments to our `Node` class to be able to contain a list of
sightings. Open up your `Node` class in Visual Studio and make the following amendments:

```
internal class Node
{
// ...

public List<Sighting> Sightings = new List<Sighting>();

public void AddSighting(Sighting sighting)
```

```
    {
    Sightings.Add(sighting);
    }

    // ...
    }
```

We will now move onto implementing `FrameGrabber`; as in the last chapter, this class will be responsible for intersecting frames from the HoloLen's color camera and making them available for analysis. The only significant difference here is what metadata we associate with it. In this example, we want to have each frame associated with the node and position and pose of the user. Using this metadata, we can create `Sighting`, as we saw before. Let's start by creating a class that will be used to store the details of a captured frame, including the required metadata. Within Visual Studio, create a new class called `Frame` in the `Content` folder and make the following amendments:

```
using System.Numerics;

internal class Frame
{
 public byte[] frameData;
 public Node node;
 public long timestamp;

 public Vector3 position;
 public Vector3 forward;
 public Vector3 up;

 public bool IsSimilar(Frame frame)
 {
   if (frame.node != node)
 {
   return false;
 }
 var dot = Vector3.Dot(forward, frame.forward);
 return dot >= 0.5f;
 }
}
```

Like the `Sighting` class, its role is to store data; in this case, we are storing the captured frame as a byte array (`frameData`) along with the associated node and the user's spatial details. We have also added the method `IsSimilar`. As mentioned previously, we want to avoid analyzing frames that are too similar--here, what we define as similar are those frames captured that share the same node (`frame.node != node`) and face in the same direction.

We determine this by calculating the dot product, which gives us a scalar of how close the forward directions of both the frames are; if the dot product of the two frames forward vectors were 1, then they would be facing in exactly the same direction. If the dot product returned was 0, then the forward directions of the frames would be perpendicular to each other.

FrameGrabber

Let's now turn our attention to the class that will be responsible for capturing the frames, most of which should look familiar to you (if you have read the previous chapter). There is a bit of code, most of which is boilerplate, so let's build it up bit by bit.

Create a new C# class in the `Content` folder, calling it `FrameGrabber`, and add the following namespaces:

```
using Windows.Graphics.Imaging;
using Windows.Media;
using System.IO;
using Windows.Storage.Streams;
using System.Numerics;
using Windows.Media.Capture.Frames;
using Windows.Media.Capture;
```

As we saw in the preceding code snippet, `FrameGrabber` is responsible for capturing frames from the color camera of the HoloLens and attaching metadata, such as the user's current position and pose. We will achieve this by passing a datasource to `FrameGrabber`, which can be called upon when creating a new frame. Create the interface for this datasource that will act as a contract between `FrameGrabber` and any object that wants to act as its data source, as in the following snippet:

```
public interface IFrameGrabberDataSource
{
Node CurrentNode { get; }
Vector3 NodePosition { get; }
Vector3 GazeForward { get; }
Vector3 GazeUp { get; }
}
```

The datasource will make available all the additional data we require for our frame (beyond the image itself). Next, let's declare all the variables and properties for our `FrameGrabber` class:

```
IFrameGrabberDataSource datasource;

MediaCapture mediaCapture;
MediaFrameSource mediaFrameSource;
MediaFrameReader mediaFrameReader;

private bool _analyzingFrame = false;
public bool IsAnalyzingFrame
{
get{
lock (this){ return _analyzingFrame; }
}
set{
lock (this){ _analyzingFrame = value; }
}
}

private bool _isNewFrameAvailable = false;
public bool IsNewFrameAvailable
{
get{
lock (this){ return _isNewFrameAvailable; }
}
set{
lock (this){ _isNewFrameAvailable = value; }
}
}

private bool _currentFrame;
public bool CurrentFrame
{
get{
lock (this){ IsNewFrameAvailable = false; return _currentFrame; }
}
set{
lock (this){ IsNewFrameAvailable = true; _currentFrame = value; }
}
}
```

Most should look familiar from the last chapter; broadly, our variables can be grouped into those related specifically to the media (mediaCapture, mediaFrameSource, and so on) and those related to managing the process pipeline (IsNewFrameAvailable, CurrentFrame, and so on). Next, we will implement the construction for FrameGrabber. Add the following constructor to the FrameGrabber class:

```
private FrameGrabber(IFrameGrabberDataSource datasource = null,
MediaCapture mediaCapture = null, MediaFrameSource mediaFrameSource = null,
MediaFrameReader mediaFrameReader = null)
{
this.datasource = datasource;
this.mediaCapture = mediaCapture;
this.mediaFrameSource = mediaFrameSource;
this.mediaFrameReader = mediaFrameReader;

if (this.mediaFrameReader != null)
{
this.mediaFrameReader.FrameArrived += MediaFrameReader_FrameArrived;
}
}
```

As we have done in the previous chapter, our construction is private, requiring FrameGrabber to be instantiated indirectly via a Factory method. The purpose of this Factory method is to hide some of the details required to construct the associated media classes, which we will see next. Add the following static method to the FrameGrabber class:

```
public static async Task<FrameGrabber> CreateAsync(IFrameGrabberDataSource
datasource)
{
MediaCapture mediaCapture = null;
MediaFrameReader mediaFrameReader = null;

MediaFrameSourceGroup selectedGroup = null;
MediaFrameSourceInfo selectedSourceInfo = null;

var groups = await MediaFrameSourceGroup.FindAllAsync();
foreach (MediaFrameSourceGroup sourceGroup in groups)
{
foreach (MediaFrameSourceInfo sourceInfo in sourceGroup.SourceInfos)
{
if (sourceInfo.SourceKind == MediaFrameSourceKind.Color)
{
selectedSourceInfo = sourceInfo;
break;
}
```

```
        }

    if (selectedSourceInfo != null)
    {
    selectedGroup = sourceGroup;
    break;
    }
    }

    if (selectedGroup == null || selectedSourceInfo == null)
    {
    return new FrameGrabber();
    }

    var settings = new MediaCaptureInitializationSettings
    {
    SourceGroup = selectedGroup,
    SharingMode = MediaCaptureSharingMode.SharedReadOnly,
    StreamingCaptureMode = StreamingCaptureMode.Video,
    MemoryPreference = MediaCaptureMemoryPreference.Cpu,
    }

    mediaCapture = new MediaCapture();

    try{
    await mediaCapture.InitializeAsync(settings);
    }
    catch(Exception e){
    return new FrameGrabber();
    }

    MediaFrameSource selectedSource =
    mediaCapture.FrameSources[selectedSourceInfo.Id];
    mediaFrameReader = await
    mediaCapture.CreateFrameReaderAsync(selectedSource);

    MediaFrameReaderStartStatus status = await mediaFrameReader.StartAsync();

    if (status == MediaFrameReaderStartStatus.Success)
    {
    return new FrameGrabber(datasource, mediaCapture, selectedSource,
    mediaFrameReader);
    }
    else{
    return new FrameGrabber();
    }
    }
```

The details of this method are described in the previous chapter, so we don't delve into them here. At a high level, we search for a suitable media source and, once found, create and initialize the `MediaCapture` instance and, then, create `MediaFrameReader`, which can be used to intercept the captured frames.

We will next implement the delegate responsible for handling frames received from `MediaFrameReader`. Add the following method to the `FrameGrabber` class:

```
void MediaFrameReader_FrameArrived(MediaFrameReader sender,
MediaFrameArrivedEventArgs args)
{
if (!IsAnalyzingFrame && !IsNewFrameAvailable)
{
MediaFrameReference frame = sender.TryAcquireLatestFrame();

if (frame != null)
{
new Task(() => SetFrame(frame)).Start();
}
}
}
```

Before processing the received frame, we check whether we are not currently analyzing a frame and whether the previous frame we processed has been consumed (`!IsAnalyzingFrame && !IsNewFrameAvailable`). If these conditions are satisfied, then we proceed with trying to acquire the last frame of `MediaFrameReader` and, if successful, passing it to the `SetFrame` method on a separate thread. The `SetFrame` method makes up the bulk of this classes' logic. It is responsible for getting the raw data from `MediaFrameReference` and pulling through the user's current position and orientation to create a new `Frame` instance. Let's implement this method now. Add the following to the `FrameGrabber` class:

```
async void SetFrame(MediaFrameReference frame)
{
IsAnalyzingFrame = true;

var node = datasource.CurrentNode;
var position = datasource.NodePosition;
var forward = datasource.GazeForward;
var up = datasource.GazeUp;

var timestamp = Utils.GetCurrentUnixTimestampMillis();

byte[] frameData = null;
try{ frameData = await GetFrameData(frame); } catch{}
if(frameData == null)
```

```
{
IsAnalyzingFrame = false;
return;
}

CurrentFrame = new Frame
{
frameData = frameData,
node = node,
position = position,
forward = forward,
up = up,
timestamp = timestamp
}
IsAnalyzingFrame = false;
}
```

As mentioned before, `Frame` consists of the raw data of the captured frame and the current state of the user in terms of their spatial information, which we obtain via assigned `datasource`. After obtaining the necessary parts, we will create a new `Frame` instance and assign it to the `CurrentFrame` property, which will flag `IsNewFrameAvailable`.

One final method to implement is, `GetFrameData`. This method takes `MediaFrameReference` and returns its video data as a byte array. Let's implement this method now; because there is a lot going on in this method, we will deliver in chunks to make it more consumable. Start by adding the following method to the `FrameGrabber` class:

```
async Task<byte[]> GetFrameData(MediaFrameReference frame)
{
byte[] bytes = null;

VideoMediaFrame videoMediaFrame = frame.VideoMediaFrame;

VideoFrame videoFrame = videoMediaFrame.GetVideoFrame();
SoftwareBitmap softwareBitmap = videoFrame.SoftwareBitmap;

SoftwareBitmap bitmapBGRA8 = SoftwareBitmap.Convert(softwareBitmap,
BitmapPixelFormat.Bgra8, BitmapAlphaMode.Ignore);

}
```

Here and throughout this book, we have omitted the majority of error checking and handling to help make the code more digestible (and save paper). Obviously, in a commercially scoped project, you would be required to check and handle the majority of the edge cases.

In the preceding snippet, we obtained the raw bitmap of the video frame associated with the media frame. We then convert this into a pixel format of **BGRA channels with 8 bytes (Bgra8)** for each. We next create a JPEG encoder and write it into an in-memory stream before reading it back out to a byte array and then, finally, returning it; amend the GetFrameData method with the following code:

```
using (InMemoryRandomAccessStream stream = new
InMemoryRandomAccessStream())
{
BitmapEncoder encoder = await
BitmapEncoder.CreateAsync(BitmapEncoder.JpegEncoderId, stream);
encoder.SetSoftwareBitmap(bitmapBGRA8);
encoder.IsThumbnailGenerated = false;

try{
await encoder.FlushAsync();
bytes = new byte[stream.Size];
await stream.AsStream().ReadAsync(bytes, 0, bytes.Length);
}
catch{}
}

return bytes;
```

This now concludes our FrameGrabber class. Let's put it to use. In the next section, we will return to our AssistantItemFinderMain class and start capturing frames using FrameGrabber.

Capturing frames

Armed with FrameGrabber, let's return to our AssistantItemFinderMain class and put it to use. Start by opening up the AssistantItemFinderMain class in Visual Studio and add the variable:

```
FrameGrabber frameGrabber;
```

Next, make the following amendments to the `SetHolographicSpace` method:

```
public void SetHolographicSpace(HolographicSpace holographicSpace)
{
// ...
InitServices();
// ...
}

async void InitServices()
{
frameGrabber = await FrameGrabber.CreateAsync(this);
}
```

If you remember, `FrameGrabber.CreateAsync` is excepting a class conforming to the interface `IFrameGrabberDataSource`. Make the necessary update (adding the `IFrameGrabberDataSource` interface to the `AssistantItemFinderMain` class). Conveniently, our `AssistantItemFinderMain` class already conforms to the `IFrameGrabberDataSource` interface; therefore, this is the only change required here.

Now, it's time to process frames; we will delegate the processing of frames to a separate method that we will poll within the `Update` method. Let's now make this amendment to `Update`:

```
public HolographicFrame Update()
{
// ...

CreateEntitiesForAllNodes(referenceFrameCoordinateSystem);

ProcessNextFrame();

timer.Tick(() =>
{
// ...
}
}
```

The task of `ProcessNextFrame` is to grab the next available frame from `FrameGrabber` and pass it off to `FrameAnalyzer` (we will be implementing this next). Before doing so, we will first check whether both `FrameGrabber` and `FrameAnalyzer` are in a valid state to proceed with the next frame, and, in addition, check whether we don't already have a frame that is similar. Similar here means two frames that would capture the same information; therefore, processing both would be redundant.

Once again, we will develop using a top-down approach, implementing the
ProcessNextFrame method and then the dependencies. Add the following to the
AssistantItemFinderMain class:

```
bool ProcessNextFrame()
{
if (!IsReadyToProcessFrame()) return false;

Frame frame = frameGrabber.CurrentFrame;
if (frame.frameData == null) return false;

if (IsFrameUnique(frame))
{
// TODO: ANALYZE FRAME
}

return true;
}
```

As mentioned in the preceding snippet, before passing the frame to Analyzer, we will first
check whether we are ready to process the next frame (IsReadyToProcessFrame) and
then, after obtaining the next frame, ensure that the frame is unique (IsFrameUnique). Let's
now implement these supporting methods, starting with IsReadyToProcessFrame:

```
bool IsReadyToProcessFrame()
{
 return frameGrabber != null &&  frameGrabber.IsNewFrameAvailable;
}
```

Here, we simply check whether frameGrabber has been initialized and we have a new
frame available. Let's now add the method IsFrameUnique to the
AssistantItemFinderMain class:

```
bool IsFrameUnique(Frame frame)
{
foreach (var node in nodes)
{
if (node != frame.node) continue;

foreach (var sighting in node.Sightings)
{
var dot = Vector3.Dot(sighting.Forward, frame.forward);
if (dot >= 0.5) return false;
}
}

return true;
```

```
    }
```

To ensure that a captured `Frame` is unique, we iterate over all the nodes and, for all the nodes that share the same `SpatialAnchor`, calculate the angle between each sighting forward direction with the frames forward direction. If we find one where the angle is minimal, that is, facing in approximately the same direction, we fail the test and ignore this frame. This now finishes off `FrameGrabber`. Let's now turn our attention to `FrameAnalyzer`.

FrameAnalyzer

Our application can now see what the user sees, but, currently, we have no way of understanding what the user is looking at. We will resolve this in this section, where we will implement our `FrameAnalyzer` class, which will pass our captured frames to Microsoft's Cognitive Vision API Service to extract the tags and caption.

Let's start by creating a new class. Within Visual Studio, create a new **C# Class** in the `Content` folder named `FrameAnalyzer`. At the top of the newly created class, add the following namespaces to get access to Microsofts Cognitive Services SDK:

```
using System.IO;
using System.Threading;
using Microsoft.ProjectOxford.Vision;
using Microsoft.ProjectOxford.Vision.Contract;
```

As we did with `FrameGrabber`, we will build this class bit by bit, starting with the variables:

```
public delegate void AnalyzedFrame(Frame frame, string description,
List<string> tokens);
public event AnalyzedFrame OnAnalyzedFrame = delegate { };
```

We first declare and delegate an event; this is how the result will be communicated back to the caller (`AssistantItemFinderMain`):

```
public Queue<Frame> frameQueue = new Queue<Frame>();
VisionServiceClient visionClient;
private string serviceKey = string.Empty;
```

We next define a queue that will be used to store the captured frames ready for processing, along with `VisionServiceClient`, which does all the heavy lifting, and a variable to store the services key that we will need to obtain from the `cognitive_service.key` file:

```
long lastFrameAnalysisTimestamp = 0;
public long frameAnalysisFrequencyMS = 6 * 1000;
```

The next set of variables will be used to throttle the service. This is to adhere to the restrictions put in place of the service:

```
ManualResetEvent processingManualResetEvent;
CancellationTokenSource processingTokenSource;
Task processingTask;
```

Our last block of variables are concerned with the task we will create to process the queue of frames.

With our variables now declared, let's implement the constructor and factory methods. Similar to `FrameGrabber`, we delegate the creation of `FrameAnalyzer` to an asynchronous creation method to hide the details:

```
public static async Task<FrameAnalyzer> CreateAsync()
{
var fa = new FrameAnalyzer();
await LoadKey(fa);

fa.visionClient = new VisionServiceClient(fa.serviceKey);
return fa;
}

async static Task LoadKey(FrameAnalyzer fa)
{
var packageFolder =
Windows.ApplicationModel.Package.Current.InstalledLocation;
var sFile = await packageFolder.GetFileAsync("cognitive_service.key");
var key = await Windows.Storage.FileIO.ReadTextAsync(sFile);
fa.serviceKey = key;
}
```

The only notable work done here is loading and reading `cognitive_service.key` for the service key and the instantiation of `VisionServiceClient`, which conveniently exposes the services of the Vision API for us.

As mentioned in the preceding snippet, this class runs on a separate thread that will continuously process the queue when a frame is available. This task is explicitly started and stopped via the Start and Stop methods, respectively. Let's add these methods now:

```
public void Start()
{
if (processingTokenSource == null)
{
processingTokenSource = new CancellationTokenSource();
}

processingTask = Task.Factory.StartNew(RunAnalyzer,
processingTokenSource.Token);
}

public void Stop()
{
if (processingTokenSource != null)
{
processingTokenSource.Cancel();
}
}
```

Nothing much of interest is happening here; within the Start method, we wrap the method RunAnalyzer in a Task to run on a separate thread. Let's keep moving forward and implement the RunAnalyzer method:

```
async void RunAnalyzer()
{
processingManualResetEvent = new ManualResetEvent(false);
while (!processingTokenSource.IsCancellationRequested)
{
processingManualResetEvent.Reset();
while (frameQueue.Count > 0)
{
// TODO: CHECK ELAPSED TIME SINCE LAST FRAME
// TODO: ANALYZE FRAME
}
processingManualResetEvent.WaitOne();
}
}
```

To make this method more interpretable, we will approach it in chunks, starting with the main loop (as seen in the preceding snippet). This method lives until `processingTokenSource` is cancelled (set via the `Stop` method) and continuously loops, processing the frames added to the queue. To make it a little more efficient, the thread is blocked until a new frame is added to the queue (`processingManualResetEvent.WaitOne()`).

Next substitute `// TODO: CHECK ELAPSED TIME SINCE LAST FRAME` with the following snippet:

```
int surplusTime = (int)(frameAnalysisFrequencyMS -
(Utils.GetCurrentUnixTimestampMillis() - lastFrameAnalysisTimestamp));

if (lastFrameAnalysisTimestamp > 0 && surplusTime > 0)
{
await Task.Delay(surplusTime);
}
```

As mentioned before, the service restricts the frequency that you can poll the service. To adhere to this, we throttle our loop by sleeping if we are processing frames too quickly.

Finally, for this method, substitute `// TODO: ANALYZE FRAME` with the following snippet:

```
var frame = frameQueue.Dequeue();

await AnalyzeFrame(frame.frameData, (status, description, tokens) =>
{
lastFrameAnalysisTimestamp = Utils.GetCurrentUnixTimestampMillis();
if (status > 0)
{
OnAnalyzedFrame(frame, description, tokens);
}
}
```

The preceding block encapsulates the main task of this class, that is, taking the next frame and passing it to the service and then broadcasting the results via the event `OnAnalyzedFrame` when the service responds. The job of analyzing the frame is delegated to the method `AnalyzeFrame`, which takes the raw image data and calls a delegate once finished and, if successful, passes the description and tokens for us to broadcast to the observers. Let's implement this `AnalyzeFrame` method. Add the following method to the `FrameAnalyzer` class:

```
async Task AnalyzeFrame(byte[] data, Action<int, string, List<string>>
callback)
{
const float minConfidence = 0.7f;
```

```
var description = string.Empty;
var tags = new HashSet<string>();
using (var stream = new MemoryStream(data))
{
VisualFeature[] visualFeatures = new VisualFeature[] {
VisualFeature.Categories, VisualFeature.Description, VisualFeature.Faces,
VisualFeature.ImageType, VisualFeature.Tags };
AnalysisResult analysisResult = await
visionClient.AnalyzeImageAsync(stream, visualFeatures);

// TODO: EXTRACT THE DESCRIPTION
// TODO: EXTRACT THE TOKENS
}

callback(string.IsNullOrEmpty(description) ? -1 : 1, description,
tags.ToList());
}
```

As we did before, we will present it in chunks to make it easier to consume. In the preceding snippet, we set a minimum confidence to 0.7. This is used when constructing the results; each result returned by the Vision API Service includes a confidence value (1.0 is equivalent to 100% confidence). This is a good indicator of to the extent to which you can trust the results. Setting this too low will mean you will include a lot of false positives, for example, cats that are not cats, and too high will mean you may miss true positives.

The service allows you to specify the analysis you want performed on the image. We define this in the line VisualFeature[] visualFeatures = new VisualFeature[] { VisualFeature.Categories, VisualFeature.Description, VisualFeature.Faces, VisualFeature.ImageType, VisualFeature.Tags }; the following list (list taken from https://westus.dev.cognitive.microsoft.com/docs/ services/56f91f2d778daf23d8ec6739/operations/56f91f2e778daf14a499e1fa) specifies the features available, but it's worth noting that each feature incurs a cost in terms of computation and, therefore, time, so it only request what is required for your task:

- **Categories**: Categorizes image content according to a taxonomy defined in the documentation.
- **Tags**: Tags the image with a detailed list of words related to the image content.
- **Description**: Describes the image content with a complete English sentence.
- **Faces**: Detects if faces are present. If present, generates coordinates, gender, and age.
- **ImageType**: Detects if image is clipart or a line drawing.

- **Color**: Determines the accent color, dominant color, and whether an image is black and white.
- **Adult**: Detects if the image is pornographic in nature (depicts nudity or a sex act). Sexually suggestive content is also detected.

`AnalysisResult` provides us with a structured way of obtaining the results. Next, we will substitute `// TODO: EXTRACT THE DESCRIPTION` and `// TODO: EXTRACT THE TOKENS`. The following shows exactly how this is done:

```
if (analysisResult.Description != null &&
analysisResult.Description.Captions != null &&
analysisResult.Description.Captions.Length > 0)
{
var caption = analysisResult.Description.Captions.OrderByDescending(x =>
x.Confidence).First();
if (caption.Confidence >= minConfidence)
{
description = caption.Text;
}

if (analysisResult.Description.Tags != null){
foreach (var tag in analysisResult.Description.Tags)
{
tags.Add(tag.ToLower());
}
}
}

if (analysisResult.Tags != null)
{
foreach (var tag in analysisResult.Tags)
{
if (tag.Confidence >= minConfidence)
{
tags.Add(tag.Name.ToLower());
}
}
}
```

We are almost finished with `FrameAnalyzer`; the last method will add new frames to the queue and signal that a new frame has arrived. Add the following method to `FrameAnalyzer`:

```
public void AddFrame(Frame frame)
{
frameQueue.Enqueue(frame);
```

```
if (processingManualResetEvent != null)
{
processingManualResetEvent.Set();
}
}
```

And this completes FrameAnalyzer; like FrameGrabber, our next task is to integrate it into our application. Let's do that now. Head back to the AssistantItemFinderMain class to make use of FrameAnalyzer.

Integrating the FrameAnalyzer

With the class AssistantItemFinderMain open in Visual Studio, let's add the variable and amend the methods responsible for the construction and destruction of FrameAnazlyer:

```
FrameAnalyzer frameAnalyzer;

async void InitServices()
{
// ...

frameAnalyzer = await FrameAnalyzer.CreateAsync();
frameAnalyzer.Start();
frameAnalyzer.OnAnalyzedFrame += FrameAnalyzer_OnAnalyzedFrame;
}

public void Dispose()
{
// ...

if (frameAnalyzer != null)
{
frameAnalyzer.Stop();
frameAnalyzer = null;
}
}
```

As you may recall, `ProcessNextFrame` was the method responsible for passing the next available frame from `FrameGrabber` and adding it to the queue of `FrameAnalyzer`. Before this method proceeded, it checked whether it was in a valid state via the `IsReadyToProcessFrame` method. Then, it would only add a frame to the queue, if it wasn't similar to an existing frame, by validating it with the method `IsFrameUnique`. Let's make amendments to each of these methods to take into account `FrameAnalzyer`, starting with the `IsReadyToProcessFrame` method:

```
bool IsReadyToProcessFrame()
{
return frameGrabber != null && frameGrabber.IsNewFrameAvailable &&
frameAnalyzer != null;
}

bool IsFrameUnique(Frame frame)
{
foreach (var queuedFrame in frameAnalyzer.frameQueue)
{
if (queuedFrame.IsSimilar(frame))
{
 return false;
}
}

// ...
}
```

Within `IsReadyToProcessFrame`, we simply extend our check, ensuring that we have an instance of `FrameAnalyzer` available. And the `IsFrameUnique` search is extended to include frames already queued up for analysis. Now, return to the `ProcessNextFrame` method and substitute `// TODO: ANALYZE FRAME` with the following:

```
frameAnalyzer.AddFrame(frame);
```

Our final task to integrate `FrameAnalyzer` is to implement the delegate `FrameAnalyzer_OnAnalyzedFrame`. Let's do that now. Add the following snippet to the `AssistantItemFinderMain` class:

```
void FrameAnalyzer_OnAnalyzedFrame(Frame frame, string description,
List<string> tokens)
{
 if (string.IsNullOrEmpty(description)) return;

 frame.node.AddSighting(new Sighting(description, frame.position,
 frame.forward, frame.up, tokens.ToArray()));
}
```

We first exit early if no description is available. Otherwise, we add a new `Sighting` to the frame's node, passing in the user's position and pose at the time when the frame was captured, along with the returned tokens and description.

This now concludes the integration of our `FrameAnalyzer`; our application now creates a trail as the user walks around and records salient objects the user sees. Our last task is, allow the user, request assistance in finding an item; but before we add this, we will build the necessary methods for constructing a path.

Constructing a path

In this section, we will implement the supporting methods that will be used to construct a path to show the user where a specific item is when requested. All of this will reside in the `AssistantItemFinderMain` class, so open this up in Visual Studio, if not already open, and let's get started.

As usual, we will start by declaring the necessary variables. Add the following to the `AssistantItemFinderMain` class:

```
private string requestedSightingTerm = string.Empty;

private Node targetNode = null;
private Sighting targetSighting = null;

private Renderer targetNodeRenderer;
```

The variable `requestedSightingTerm` stores the item we are currently searching for, while `targetNode` and `targetSighting` will be the associated node and sighting related to this search term. `targetNodeRenderer` will be used to differentiate the last node from the others. Let's start by instantiating `targetNodeRenderer`. Make the following amendments to the `AssistantItemFinderMain` class:

```
public void SetHolographicSpace(HolographicSpace holographicSpace)
{
// ...

nodeRenderer = new NodeRenderer(deviceResources, new Vector3(0.3f, 0.3f,
1.0f), 0.05f);
nodeRenderer.CreateDeviceDependentResourcesAsync();

targetNodeRenderer = new NodeRenderer(deviceResources, new Vector3(0.3f,
1.0f, 0.3f), 0.09f);
targetNodeRenderer.CreateDeviceDependentResourcesAsync();

// ...
```

```
}

public void Dispose()
{
// ...

if (nodeRenderer != null)
{
nodeRenderer.Dispose();
nodeRenderer = null;
}

if (targetNodeRenderer != null)
{
targetNodeRenderer.Dispose();
targetNodeRenderer = null;
}

// ...
}
public void OnDeviceLost(Object sender, EventArgs e)
{
// ...
nodeRenderer.ReleaseDeviceDependentResources();
targetNodeRenderer.ReleaseDeviceDependentResources();
// ...
}
public void OnDeviceRestored(Object sender, EventArgs e)
{
// ...
nodeRenderer.CreateDeviceDependentResourcesAsync();
targetNodeRenderer.CreateDeviceDependentResourcesAsync();
// ...
}
```

With the management of `targetNodeRenderer` now complete, let's turn our attention to the `Update` method. Currently, we are creating an entity for each node. We want to change this, so that we only create entities (visual representations of the nodes) when we are constructing a path to `targetNode`, that is, when we are assisting the user to a specified item.

Jump into the `Update` method where we start making these changes, starting with removing the statement `CreateEntitiesForAllNodes(referenceFrameCoordinateSystem)`.

Our first addition here has to do with setting `targetNode` and `targetSighting`. This will be performed when the variable `requestedSightingTerm` is set (setting this will be the topic of the next section). Add the following block to the `Update` method:

```
if (!string.IsNullOrEmpty(requestedSightingTerm))
{
var candidates =
FindClosestNodesWithSightedItem(referenceFrameCoordinateSystem, pose,
requestedSightingTerm);

if (candidates != null && candidates.Count > 0)
{
targetNode = candidates[0];
targetSighting = candidates[0].Sightings.Where(sighting =>
sighting.Tokens.Any(token => token.Equals(requestedSightingTerm,
StringComparison.OrdinalIgnoreCase))).First();
}
requestedSightingTerm = string.Empty;
}
```

When `requestedSightingTerm` is not null, we search for the closest node (relative to the user) for `Sighting` that has a tag equal to `requestedSightingTerm`. If found, we set `targetNode` and `targetSighting`. Searching is delegated to the method `FindClosestNodesWithSightedItem`, which we will return to very soon.

Our next block of code is concerned with dismissing a target. This occurs when the user is in proximity (same node) of `targetNode` and has resided there longer than a set period. Add the following code just under the preceding snippet:

```
if (CurrentNode == targetNode)
{
if (dwellTimeAtCurrentNode >= 5)
{
targetNode = null;
targetSighting = null;
entities.Clear();
}
}
```

The final addition to the `Update` method is the construction of the path if `targetNode` is set. Add the following snippet:

```
if (targetNode != null)
{
  RebuildTrailToTarget(referenceFrameCoordinateSystem, prediction.Timestamp,
CurrentNode, targetNode);
}
```

Here, we simply delegate the construction of the path to the method `RebuildTrailToTarget`. Our `Update` method is now complete, ignoring the errors. Let's now move on to implementing some of the dependencies, namely, `FindClosestNodesWithSightedItem` and `RebuildTrailToTarget`.

Let's start with `FindClosestNodesWithSightedItem`. This method simply constructs and returns a list of nodes, filtering out those that don't have a sighting with the specified tag (`sightingItem`) and then orders them based on their distance from the user:

```
public IList<Node> FindClosestNodesWithSightedItem(SpatialCoordinateSystem
referenceFrameCoordinateSystem, SpatialPointerPose pose, string
sightingItem)
{
var filteredNodes = nodes.Where(node =>
{
return node.Sightings.Any(sighting =>
{
return sighting.Tokens.Any(token => token.Equals(sightingItem,
StringComparison.OrdinalIgnoreCase));
});
});

if (filteredNodes != null)
{
return filteredNodes.OrderBy(node =>
{
  return node.TryGetDistance(referenceFrameCoordinateSystem,
pose.Head.Position);
});
}
return null;
}
```

Our next method, `RebuildTrailToTarget`, replaces our `CreateEntitiesForAllNodes`--instead of creating an entity for all nodes, it finds the shortest path, based on the edges connecting the nodes, from the given start and end nodes, and then creates entities for each of the nodes along this path. Add the following method to the `AssistantItemFinderMain` class:

```
int RebuildTrailToTarget(SpatialCoordinateSystem
referenceFrameCoordinateSystem, PerceptionTimestamp perceptionTimestamp,
Node startNode, Node endNode,
            int lookAhead = 100)
{
entities.Clear();

Stack<Node> trail = new Stack<Node>();
BuildPath(endNode, startNode, trail);

if (trail.Count == 0) return -1;

var rootEntity = GetRootEntity(referenceFrameCoordinateSystem);
int i = 0;
while (i < lookAhead && trail.Count > 0)
{
var node = trail.Pop();

var entity = new Entity($"node_{i}");
entity.Node = node;
entity.Renderer = entity.Node == targetNode ? targetNodeRenderer :
nodeRenderer;
entity.UpdateTransform(referenceFrameCoordinateSystem);
var targetPosition = rootEntity.Transform.Translation;
entity.Position = new Vector3(0, (targetPosition -
entity.Transform.Translation).Y, 0f);
entities.Add(entity);

i = i + 1;
}

return i;
}
```

We first clear the `entities` collection and then delegate the path planning to the method `BuildPath`, passing it through the start and end nodes along with a collection to store the path--here, to construct the path from the node where the item is located back to the user. This reduces our search space and, therefore, improves the performance.

Once the path has been constructed, we create a root entity (`rootEntity`), as we did before, to use as a reference for the *y* position for all the entities. Then, we iterate through the collection of nodes and create an associated entity, almost the same as we did in `CreateEntitiesForAllNodes` with the only difference, that we assign `targetNodeRenderer` to the end node and `nodeRenderer` to all the others. This is used to help communicate to the user that this is the location of the item they are searching for.

Let's finish off the construction of the path by implementing the last method, `BuildPath`. `BuildPath` is a recursive function that traverses the connected nodes (using the edges) until it finds the node where the user resides (remembering that we are searching from the Node where the item is to the user), or until a dead end, that is, there are no other nodes to traverse. When this node is found, it builds up the path while it unwinds up the stack from the recursive calls. Let's see what this looks like in code. Add the following snippet to the `AssitantItemFinderMain` class:

```
bool BuildPath(Node currentNode, Node endNode, Stack<Node> trail)
{
List<Node> connectedNodes = edges.Where(edge => edge.NodeA == currentNode
|| edge.NodeB == currentNode).Select(edge => (edge.NodeA != currentNode ?
edge.NodeA : edge.NodeB)).ToList();

if (connectedNodes.Contains(endNode))
{
trail.Push(currentNode);
return true;
}
else{
foreach (var node in connectedNodes)
{
if (BuildPath(node, endNode, trail))
{
trail.Push(currentNode);
return true;
}
}
}
return false;
}
```

To elaborate on the preceding explanation, we first find all the edges associated with `currentNode`. We then check for our termination condition, which is if one of the edges connects to our `endNode`. If so, then we add `currentNode` to the trail (our path) and return true. Otherwise, we iterate through each of the connected nodes and recursively call `BuildPath`, which will return true if the path leads to `endNode`, in which case, we append `currentNode` to the trail. Otherwise, we ignore.

This now completes the code required for constructing a path. We are left with setting `targetNode`. As seen before, this node is set in the `Update` method when `requestedSightingTerm`. When this is set, we enter a block that searches through each node's (starting with the closest to the user) sighting for a matching tag. If found, then we set the node to `targetNode` which, in turn, constructs our path. The next, and final, section will cover how we set the variable `requestedSightingTerm`.

Getting assistance

The last part of the functionality we need to add is giving the user the ability to get assistance finding an item. For this, we will be using--voice. In this example, we will be using a prebuilt class bundled with the starter project, but will go into the details later on in this book, when we explore different modes of interaction. The class we will be making use of is the appropriately named `SpeechManager`. This class provides a very minimalist implementation for speech recognition; for each tag we identify, we add to `SpeechManager` via the asynchronous method `AddTagsAsync`, which will create phrases using the prefixes:

- find tag
- where is tag
- locate tag

When a phrase is recognized, `SpeechManager` will broadcast the `OnPhraseRecognized` event, passing the status, text, and tag. We will only process events that have a status equal to 1, indicating that the confidence is medium or higher on the recognized phrase. We then split the tag, delimited by the character "_", where the first part indicates the intent and the second the tag (or item) we want to find.

Let's jump back into the `AssistantItemFinderMain` class to hook this all up. As we have done previously, start by adding the variable to the holder reference to an instance of `SpeechManager`.

```
private SpeechManager speechManager;
```

Next, initialise it in the `InitServices` method and handle stopping it in the `Dispose` method. Amend each of these methods with the following code:

```
async void InitServices()
{
 // ...

 speechManager = await SpeechManager.CreateAndStartAsync();
 speechManager.OnPhraseRecognized += SpeechManager_OnPhraseRecognized;
}
```

In the preceding snippet, we are creating a new instance of it and, once instantiated, assigning a delegate to the `OnPhraseRecognized` event:

```
public void Dispose()
{
 // ...

 if(speechManager != null)
 {
  speechManager.Stop();
  speechManager = null;
 }
}
```

There is only one more method left: `SpeechManager_OnPhraseRecognized` to handle events from `SpeechManager`. As mentioned in the preceding code snippet, we ignore events with a status less than one, and split the tag parameter to determine the intent and item to search for. Let's add this method to the `AssistantItemFinderMain` class:

 After some experimentation, it was discovered that Microsoft's Cognitive Vision API worked satisfactory with large salient objects in the scene (such as a bike) but wasn't picking up smaller objects (such as keys). To prototype the concept, an additional intent was added and can be used manually to place items in the environment. The following is a list of phrases you can use to manually place items: remember keys, remember wallet, remember hat, remember umbrella, remember coat, remember jacket, and remember hammer.

```
void SpeechManager_OnPhraseRecognized(int status, string text, string tag)
{
 if (status < 0) return;

 var tagSplit = tag.Split('_');
 var intent = tagSplit[0];
 var item = tagSplit[1];
```

```
if (intent.Equals("findLocation", StringComparison.OrdinalIgnoreCase))
{
 requestedSightingTerm = item;
}
else{
 var node = CurrentNode;
 var position = NodePosition;
 var forward = GazeForward;
 var up = GazeUp;

 node.AddSighting(new Sighting(text, position, forward, up, item));
 }
}
```

Once we have extracted the intent and item from the tag parameter, we test to see if we are finding the item or adding it. If finding, then we simply set the item value to the variable `requestedSightingTerm`, where the path construction of the navigation path will be handled in the `Update` method. If we are adding an item, then we instantiate a new `Sighting` and add it to `currentNode`.

This concludes this example and it's time to take it for a test run. Turn ON your HoloLens, if OFF, and click on the green arrow just in front of the **Remote Machine** button in the toolbar, as we did before, to build and deploy to the device. Once deployed and running, walk around your environment and manually place some items to test the path construction:

Summary

Congratulations! This was a lengthy chapter but I hope you enjoyed the journey and have been inspired to create new experiences that converge the digital and physical world. In this chapter, our main focus was on the coordinate system and making use of spatial anchors to navigate around a large space. This is a fundamental concept for the HoloLens, as it is one of the major differentiating factors between developing for VR and MR, and will provide valuable insights as you move forward developing your MR applications.

In the next chapter, we change gears and move onto a different toolset, Unity, where we continue exploring development for HoloLens, including spatial mapping and an in-depth look into gaze, gestures, and voice--the three main modes for interacting with MR applications.

4
Building Paper Trash Ball in Unity

Imagine that you are sitting at your desk; in front of you is a scrap piece of paper and not too far to your left is the rubbish bin (let's be environmentally conscious and make it a recycling bin). Like many of your peers, you instinctively crumble up the piece of paper and, with your best attempt, throw the crumpled piece of paper into the bin.

There's something about this simple act that is satisfying, so much so that a game created by Backflip Studios, called *Paper Toss*, remained number one on the app store for most of 2009.

To get ourselves acquaint with developing for HoloLens using Unity, we will reimagine this game for **mixed reality** (**MR**); by the end of this chapter and you will know how to do the following:

- Configuring a Unity project for HoloLens
- Mapping and understanding your environment
- Placing holograms in the environment using gaze and the tap gesture
- Allowing the user to toss the paper in the bin using gaze and gestures

There is a lot to get through, so let's jump right in.

Configuring a Unity project for HoloLens

Let's start from the top; our first task is setting up Unity and developing HoloLens applications. In this section, we will walk through setting up Unity, create a simple scene, and finally; deploy it to the emulator and device to ensure that everything is set up correctly.

At the time of writing, you will need to use Unity HoloLens Technical Preview or Unity Beta 5.5b9 or greater; you can find details at the official Unity site, `http://www.unity.com`. In addition, it would be useful to download and install the HoloLens emulator, which can be downloaded from `http://go.microsoft.com/fwlink/?LinkID=823018`. Lastly, ensure that you have Visual Studio 2015 with Update 3 installed.

Once downloaded and installed, launch Unity and create a new 3D project, as shown in the following screenshot:

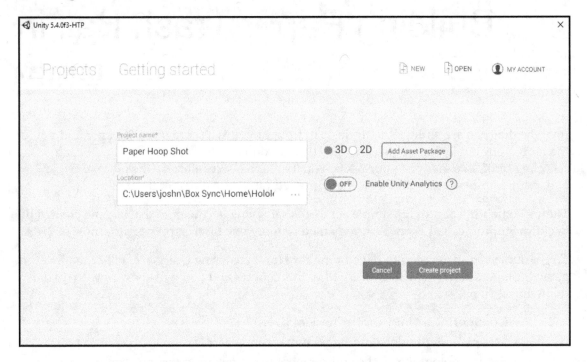

After clicking on the **Create project** button, you will be presented with an empty Unity workspace, and now the fun begins. Introducing Unity is beyond the scope of this book. I hope to cover enough for you to work through the examples within this book, but would recommend *Unity 5.x By Example* by Alan Thorn to get a more rounded and comprehensive introduction to Unity.

Start by importing the `PaperHoopShootAssets_START.unitypackage` Unity package; you can import the assets either by double-clicking on the package or within Unity via the **Assets | Import | Custom Packages** menu. This will bring up the **Import** dialog from Unity, as in the following screenshot:

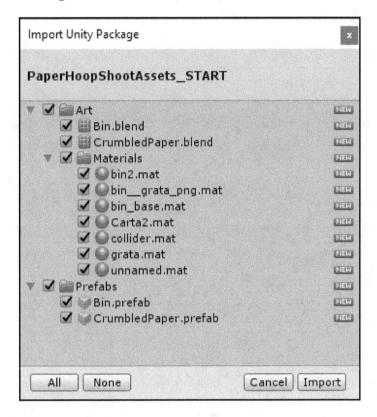

Click on **Import** to bring the assets into your project.

It's worth taking a moment to appreciate the minimal number of assets needed for our game; developing MR experiences allows us to borrow from the real world.

With our assets now imported, let's turn our attention to configuring Unity for HoloLens. Open the **Build Settings** dialog, as illustrated, via the **File** | **Build Settings** menu:

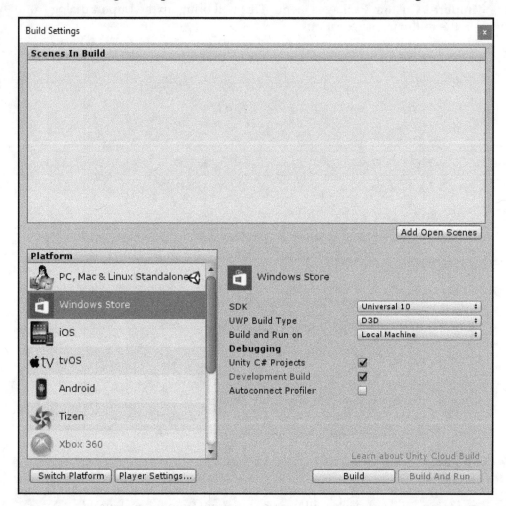

Configure the **Build Settings** to resemble what is presented in the preceding screenshot:

- Set **Platform** to **Windows Store**
- Set the **SDK** to **Universal 10**
- Set **UWP Build Type** to **D3D**
- Check **Unity C# Projects**
- Check **Development Build**

Commit these changes by clicking on the **Switch Platform** button; then, click on the **Player Settings...** button to open the **Player Settings** dialog, where we will continue configuring Unity for HoloLens.

Expand the **Other Settings** pane and make the following changes:

- Check **Use 16-bit Depth Buffers**
- Check **Virtual Reality Supported**

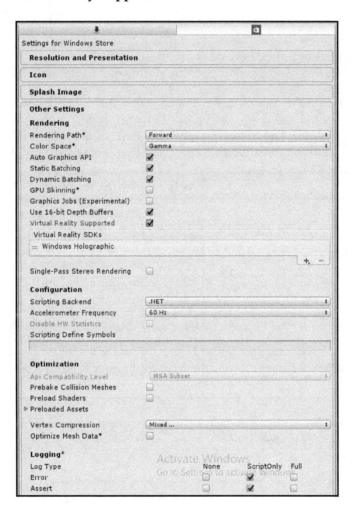

Checking **Use 16-bit Depth Buffers** forces Unity to use a 16-bit depth buffer instead of a 24-bit buffer. This reduces the amount of data pushed between the CPU and **Graphics Processing Unit** (**GPU**), which improves performance and reduces power consumption; however, because it reduces the resolution of the depth that can be captured in your scene, it can cause Z-fighting for objects that are tightly clustered together.

 Z-fighting is when two or more primitives have similar or identical values in the depth texture, causing flicking when rendered.

Checking **Virtual Reality Supported** makes the associated libraries available; once checked, you should see **Windows Holographic** become visible in a list directly below it.

We now need to declare the capabilities we need; expand the **Publisher Settings** pane of the **Player Settings** panel and check the **SpatialPerception**, **Microphone**, and **InternetClient** capabilities. The following table, taken from the official HoloLens documentation, shows the relevant capabilities and their associated APIs:

Capability	APIs requiring capability
WebCam	`PhotoCapture` and `VideoCapture`
SpatialPerception	`SurfaceObserver` and `SpatialAnchor`
Microphone	`VideoCapture`, `DictationRecognizer`, `GrammarRecognizer`, and `KeywordRecognizer`
InternetClient	`DictationRecognizer` (and to use the Unity Profiler)
PicturesLibrary/VideosLibrary /MusicLibrary	`PhotoCapture` and `VideoCapture` (for still photos, video, and audio, respectively)

As HoloLens is a mobile device, it requires the same considerations for balancing power consumption and performance.

The easiest way to ensure that you deliver an adequate experience is to update your projects setting to **Fastest**; to do this, open the **QualitySettings** dialog via **Editor** I **Project Settings** I **Quality**, and set the level to **Fastest**, as shown in the following screenshot:

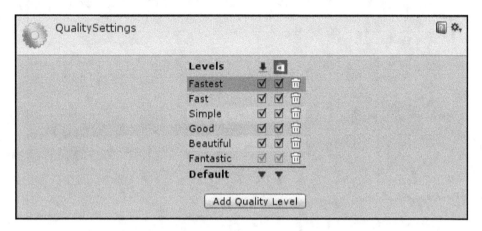

Creating our first Mixed Reality scene

HoloLens works seamlessly with Unity. With minimal changes to the camera, we can quickly design, build, and deploy MR applications; we'll build one right now (albeit a very simple one).

Select the **Main Camera** in the **Scene1** window of the **Hierarchy** panel and make the following changes in the **Inspector** panel:

1. Set the **Position** of **X** as 0, **Y** as 0, and **Z** as 0.

2. Select **Solid Color** for the **Clear Flags** field and set the **Background** color to black; black renders as transparent in HoloLens.

3. Set the **Clipping Planes** I **Near** as 0.85 and **Far** as 10. Here, the value of 0.85 is a recommendation from Microsoft as spatial mapping is currently effective between the range of 0.85 meters and 4 meters, where 1 unit in Unity represents 1 meter in the physical world.

The following screenshot shows the changes made within the **Inspector** panel:

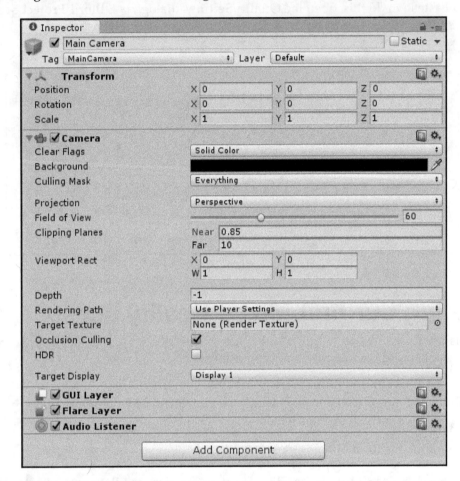

Find the **Bin** prefab in the **Project** panel within the **Prefabs** folder and drop it into the **Scene** window; select the **Bin** either by clicking on it in the **Hierarchy** panel or within the **Scene** window. Now, with that selected, update the **Position** of **X** as 0, **Y** as −0.5, and **Z** as 2 from within the **Inspector** panel. This effectively positions the bin 2 meters in front and 0.5 meters below the HoloLens device (also the position of the user's head commonly).

Prefabs in Unity are preconfigured GameObjects that act as templates; they provide a convenient way to quickly create and manage multiple instances of similar objects.

Let's see how this looks on the device; click on the **File** | **Build Settings** menu item:

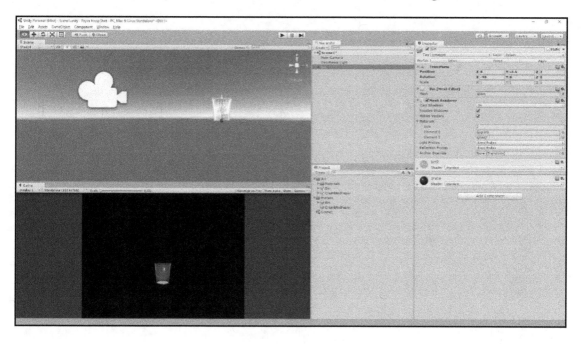

The Build Settings dialog box

This will open the **Build Settings** dialog. Click on **Add Open Scenes** to include the current scenes in **Scenes In Build** and click on the **Build** button to build the project:

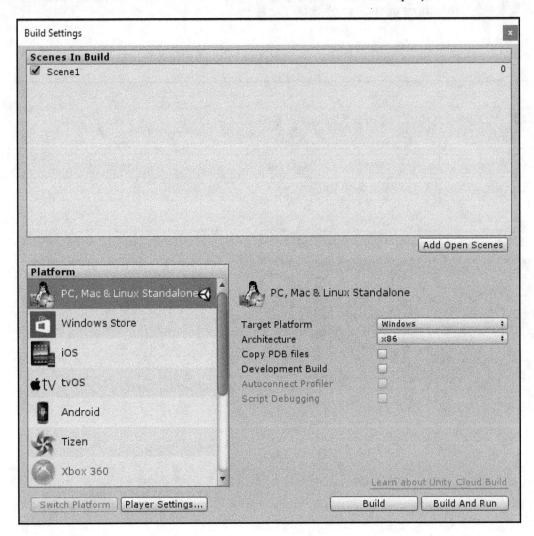

Once finished, open the root directory of the build and double-click on the solution (.sln file) to open Visual Studio, and finally, build and deploy to either the device or emulator. Refer to Chapter 9, *Deploying Your Apps to the HoloLens Device and Emulator*, for details.

We now have a holographic bin floating around our room. As impressive as this is, there is something very uncanny about it; maybe it has something to do with the fact that the bin is defying physics and floating mid-air. In the following section, we will implement the functionality to place the bin firmly onto the floor in our room.

Mapping our environment

As discussed in Chapter 1, *Enhancing Reality*, spatial mapping is the process of scanning and mapping the physical environment, making it digitally accessible. Having access to the physical environment allows us to create experiences that better integrate with the real world, by creating a greater illusion of realism or utility through understanding the user's context better. In this chapter, we are interested in creating a more immersive experience by merging the virtual and real worlds.

Here, we will learn how to scan and visualize the environment and then use this mapping to place our bin onto a suitable surface, a place where one would expect to find a bin in the real world--on the floor.

Unity provides two ways we can perform spatial mapping; the first uses prebuild components and the second uses the low-level APIs from the UnityEngine.VR.WSA namespace. In the subsequent sections, we will explore both approaches, starting with the low-level APIs.

Just before jumping into code, let's create a new layer to assign to our scanned surfaces. This provides a convenient way of filtering when performing collision checks. From the menu, select **Edit** | **Project Settings** | **Tags and Layers**.

This will open a panel, **Layers**, listing all the layers, as in the following screenshot; add `SpatialSurface` at index **User Layer 31** and `Hologram` at index **User Layer 30**; now `SpatialSurface` will be assigned to our surfaces and `Hologram` will be assigned to our holograms, or bin in this case:

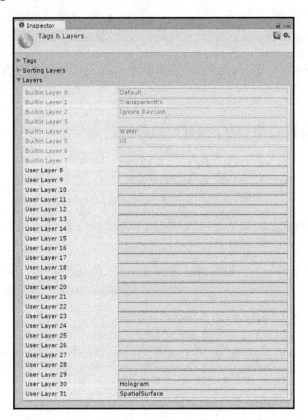

Spatial mapping API

Let's start at the bottom and work our way up. In this section, we will learn how to use low-level APIs to map and visualize our environment, and then, in the subsequent section, we will look at an alternative approach using the prebuilt components.

The class responsible for initiating and managing the mapping process is `SurfaceObserver`; this communicates its results to the registered delegates: `SurfaceDataReadyDelegate` and `SurfaceChangedDelegate`.

To start with, create a new empty GameObject via the **GameObject | Create Empty** menu item, rename the GameObject to `SpatialMappingManager` by selecting it in the **Project** window of the **Hierarchy** panel, right-clicking, and selecting the **Rename** option.

 Unity's architecture is heavily influenced by the Composite Pattern, where each entity is made up of a collection of components. This allows encapsulation across domains for a single entity. An example is a component handling rendering and another for physics. In Unity, GameObject is the host container with the mandatory `Transform` component attached, which encapsulates the positioning information.

We will create a new C# script that will be responsible for managing and interacting with the class `SurfaceObserver`. With `SpatialMappingManager` selected, click on the **Add Component** button from within the **Inspector** panel and select **New Script**; enter the name `SpatialMappingManager`. Your **Scene** should now look as shown in the following screenshot:

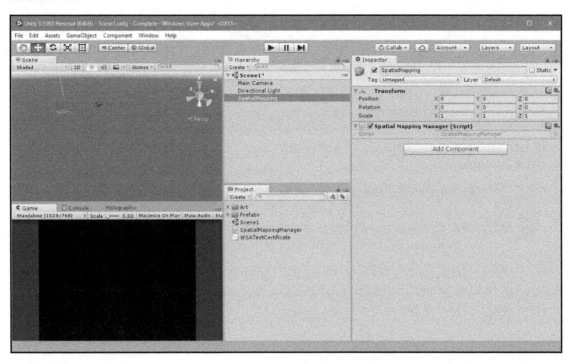

Double-click on `SpatialMappingManager` shown in the **Project** panel; this will launch Visual Studio with the file open.

If you have continued from building and deploying in the previous section, then you will likely have two instances of Visual Studio open. One instance will be associated with the build, and the other for editing Unity scripts; the latter is the instance we will be working in.

Start by including the following namespaces at the top of the script and instantiating an instance of class `SurfaceObserver` from within the `Start` method:

```
using System;
using System.Collections;
using System.Collections.Generic;
using UnityEngine;
using UnityEngine.VR.WSA;

SurfaceObserver surfaceObserver;

void Start ()
{
    surfaceObserver = new SurfaceObserver();
}
```

`SurfaceObserver` is responsible for one region of space defined by a list of bounding volumes; these bounding volumes are added to the class `SurfaceObserver` using the `SetVolumeAsAxisAlignedBox`, `SetVolumeAsFrustum`, `SetVolumeAsOrientedBox`, and `SetVolumeAsSphere` helper methods. The shape and size of your bounding volume will be specific to your specific application's requirements. When updated, the `SurfaceObserver` will notify the registered delegate of new, updated, and removed surfaces from within the specified volumes.

An application can have multiple `SurfaceObserver` classes; one reason for this can be for performance where you might have multiple `SurfaceObserver` classes with varying levels of detail, allowing you to switch level of details depending on different states of the application.

Here (in Paper Trash Ball), we are interested in the area around the user, so we will set the bounding volume to encapsulate a sphere with a radius of 2 meters; do this by adding the following to the `Awake` method:

```
void Awake ()
{
  surfaceObserver = new SurfaceObserver();
  surfaceObserver.SetVolumeAsSphere(Vector3.zero, 2.0f);
}
```

To receive updates, we must consciously poll the `SurfaceObserver`; we do this by calling the asynchronous `Update` method of the `SurfaceObserver`, passing in a `SurfaceChangedDelegate` delegate to handle the results. Let's do that now; add the following code to our `SpatialMappingManager` class:

```
public float timeBetweenUpdates = 3.0f;
bool _observing;

public bool IsObserving
{
    get { return _observing; }
    set
    {
        _observing = value;
        StopAllCoroutines();
        if (_observing)
        {
            StartCoroutine(Observe());
        }
    }
}
IEnumerator Observe()
{
    var wait = new WaitForSeconds(timeBetweenUpdates);
    while (IsObserving)
    {
        surfaceObserver.Update(onSurfaceChanged:OnSurfaceChanged);
        yield return wait;
    }
}
```

In the preceding code snippet, we have implemented the `IsObserving` property, which is responsible for starting and stopping polling; when `true`, we will start a coroutine to continuously poll the `SurfaceObserver` for updates every 3 seconds and stop if false.

> A coroutine is a type of function that can suspend execution, returning control to Unity, and then resuming where it left off in subsequent frames.

At this point, you will likely be receiving an error from Visual Studio complaining that `OnSurfaceChanged` is not implemented; let's fix that now by stubbing this method out.

Add the following method to our `SpatialMappingManager` class:

```
void OnSurfaceChanged(SurfaceId surfaceId, SurfaceChange changeType, Bounds
bounds, DateTime updateTime)
   }
```

This is our delegate method for handling results received from the `SurfaceObserver`; it is passed the `surfaceId`, `changeType`, `bounds`, and `updateTime` parameters. The parameter passed, `surfaceId`, is a unique identifier for the observed surface; `changeType` is an enumeration of the `SurfaceChange` type, describing the type of update (whether it is added, updated, or removed); `bounds` describes the bounding area of the surface; and finally, `updateTime` is a timestamp of the observation (which can be useful if you are rapidly polling the `SurfaceObserver`).

At this point, we still have no information that we can use for our application to determine where to place our bin (the ground); for this, we must explicitly request the `SurfaceObserver` to bake the surface mesh.

Create a generic `Dictionary` using `SurfaceId.handle`, which is the surface identifier for the key and GameObject for the value; for our `SpatialMappingManager` class, this `Dictionary` will hold references to all observed surfaces:

```
Dictionary<int, GameObject> cachedSurfaces = new Dictionary<int,
GameObject>();
```

Within the `OnSurfaceChanged` method, add the following switch state to handle the different types of observation:

```
void OnSurfaceChanged(SurfaceId surfaceId, SurfaceChange changeType, Bounds
bounds, DateTime updateTime)
{
switch (changeType)
        {
            case SurfaceChange.Added:
            case SurfaceChange.Updated:
                {
                    break;
                }
            case SurfaceChange.Removed:
                {
                    break;
                }
        }
    }
```

When observing a removed surface, we want to destroy the associated GameObject and remove it from our cache, as shown in the following code snippet:

```
case SurfaceChange.Removed:
{
 GameObject surface;
 if (cachedSurfaces.TryGetValue(surfaceId.handle, out surface))
 {
  cachedSurfaces.Remove(surfaceId.handle);
  GameObject.Destroy(surface);
 }
break;
}
```

The process of building and storing surface mesh data is computationally expensive, so how you handle changes is important and determined largely by your application's requirements. The following are some considerations to help you think about some possible strategies:

Use a priority queue to handle updates, giving higher priority to surfaces closest to the user.

Keep surfaces cached when the user is frequently switching between surface volumes, even if the `SurfaceObserver` returns a `Deleted` type.

Use low-and-high resolution `SurfaceObserver` for processing high-resolution surfaces only when required.

Let's improve how we handle surfaces that are removed. Our strategy will be based on some assumptions about the user, mainly that they will be mostly stationary and likely to initially search around the environment; as they are looking around, it is possible that HoloLens will remove surfaces that become out of view of the user. In our current implementation, we prematurely remove these surfaces. A better way, in this context, is to have some way of expiring surfaces once observed being removed to avoid having to recreate them. We can easily do this by adding them to a removal queue that is flushed less frequently than this update cycle. Let's go ahead and make the necessary changes. Start by declaring a data structure that will be used for removing expired surfaces, including a variable indicating the expiry time:

```
public float removalDelay = 10.0f;

SurfaceObserver surfaceObserver;

Dictionary<int, GameObject> cachedSurfaces = new Dictionary<int,
```

```
GameObject>();
Dictionary<int, float> surfacesToBeRemoved = new Dictionary<int, float>();
```

Now update the `SurfaceChange.Removed` block in the `OnSurfaceChanged` method with the following code:

```
case SurfaceChange.Removed:
{
        GameObject surface;
        if (cachedSurfaces.TryGetValue(surfaceId.handle, out surface))
        {
                surfacesToBeRemoved.Add(key: surfaceId.handle, value:
                Time.time + removalDelay);
        }
        break;
}
```

To avoid expiring surfaces that come back into view, we will remove surfaces from the `surfacesToBeRemoved` collection when we are notified of *that* surface being added or updated. Add the following code to the `SurfaceChange.Added` and `SurfaceChange.Updated` blocks of the `OnSurfaceChanged` method:

```
case SurfaceChange.Added:
case SurfaceChange.Updated:
{
        if(surfacesToBeRemoved.ContainsKey(key: surfaceId.handle))
        {
                surfacesToBeRemoved.Remove(key: surfaceId.handle);
        }

        break;
}
```

The final logic we need to implement is the removing of the surfaces once they have expired. We do this by checking whether the current time is greater than the set expiry time; add the following code to the `Update` method:

```
void Update()
{
    var surfaceIds = surfacesToBeRemoved.Keys;
    foreach(int surfaceId in surfaceIds)
    {
        if (surfacesToBeRemoved[surfaceId] >= Time.time)
        {
            surfacesToBeRemoved.Remove(surfaceId);

            GameObject surface;
```

```
                    if (cachedSurfaces.TryGetValue(surfaceId, out surface))
                    {
                        cachedSurfaces.Remove(key: surfaceId);
                        GameObject.Destroy(surface);
                    }
                }
            }
        }
```

With our surface removal strategy implemented, let's now turn our attention to implementing the functionality to handle added and updated observations.

When we observe that a surface is added or updated, we will request the latest mesh data. Add the following code to the `SurfaceChange.Added` and `SurfaceChange.Updated` blocks of our `OnSurfaceChanged` method:

```
    {
            GameObject surface;
            if (!cachedSurfaces.TryGetValue(surfaceId.handle,
            out surface))
            {
                surface = new GameObject();
                surface.name = string.Format("surface_{0}",
                surfaceId.handle);
                surface.layer =
                LayerMask.NameToLayer("SpatialSurface");
                surface.transform.parent = transform;
                surface.AddComponent<MeshRenderer>();
                cachedSurfaces.Add(surfaceId.handle, surface);
            }
    }
    break;
    }
```

We use GameObject as the container for each observed surface with the surfaces mesh bound to an attached `MeshFilter` component, which is created, if one doesn't already exist in our `cachedSurfaces` collection.

We are now ready to request the `SurfaceObserver` to bake the mesh data for the surface. We do this by first populating a `SurfaceData` struct indicating what information we require, and then passing this to the `RequestMeshAsync` method of the `SurfaceObserver`, along with our `SurfaceDataReadyDelegate` delegate. Append the following code to the `SurfaceChange.Added` and `SurfaceChange.Updated` blocks of our `OnSurfaceChanged` method:

```
            {
    . . .
```

```
var surfaceData = new SurfaceData(
                    _id: surfaceId,
                    _outputMesh: surface.GetComponent<MeshFilter>()
?? surface.AddComponent<MeshFilter>(),
                    _outputAnchor:
surface.GetComponent<WorldAnchor>() ?? surface.AddComponent<WorldAnchor>(),
                    _outputCollider:
surface.GetComponent<MeshCollider>() ??
surface.AddComponent<MeshCollider>(),
                    _trianglesPerCubicMeter: 1000,
                    _bakeCollider: true
                );
        surfaceObserver.RequestMeshAsync(dataRequest:
surfaceData, onDataReady: OnDataReady);
            break;
        }
```

The result is returned to the `SurfaceDataReadyDelegate` delegate, passing a baked surface mesh, assigned to the associated `surfacesGameObjectsMeshFilter`.

The level of detail, determined by the value assigned to `trianglesPerCubicMeter`, determines at what resolution HoloLens describes the surfaces. Higher values provide high resolution but also higher computational and storage costs; therefore, it's important to choose a resolution adequate for your application.

> For those not familiar with Unity; both the preceding code snippets introduce some new components, including `MeshRenderer`, `MeshFilter`, and `MeshCollider`.

> Here, we briefly describe what each component does.
> **MeshFilter** is the component that holds reference to the mesh data used to describe the geometric data of your model; this mesh is then used by **MeshRenderer** to render it onto a screen with the assigned material(s). The **MeshCollider** component is responsible for describing the bounds (form) of the object, which is used by the physics engine.

Setting `_bakeCollider` to `true` and passing through a reference of the `MeshCollider` component of GameObject will give us a GameObject we can easily interact with.

We also pass through a `WorldAnchor` component; this locks the surface in place relative to real-world objects and is used as a frame of reference by the application to orientate and position nearby holograms.

 HoloLens uses the "stationary frame of reference" as a reference to its world origin; this is created at startup, and holograms are positioned relative to it (this is how we previously positioned the bin). HoloLens also offers another method for anchoring holograms to the real world; in Unity, this is the `WorldAnchor` component. When a `WorldAnchor` component is attached to GameObject, HoloLens prioritizes the sensor data around that anchor. This has the advantage of more accurate tracking and the possibility of sharing the anchor with other devices so that they can reason about the same location.

 In the preceding code snippet, we attached a `WorldAnchor` to our `Surface` GameObject; you can just as easily attach a `WorldAnchor` to any GameObject, but doing so removes the ability to move the object.

Once again, Visual Studio will be complaining that `OnSurfaceChanged` is not implemented; let's fix that now by stubbing this method out, as in the following code snippet:

```
    void OnDataReady(SurfaceData bakedData, bool outputWritten, float
elapsedBakeTimeSecond)
    {

    }
```

The delegate returns the `SurfaceData` along with the `outputWritten` boolean; this indicates whether the process was successful or not as well as `elapsedBakeTimeSecond`, a value indicating how long the process took.

We've made some good progress; go ahead and build and deploy to the emulator or device to see how far we have come--you should see something like the following:

The mesh is being rendered using the default wireframe *Shader*; the colors represent the distance from the origin. Here's the list of colors and the associated distances:

- 0 - 1 meter = Black
- 1 - 2 meters = Red
- 2 - 3 meters = Green
- 3 - 4 meters = Blue
- 4 - 5 meters = Yellow
- 5 - 6 meters = Cyan
- 6 - 7 meters = Magenta

Pretty, but not for everyone--let's make some minor additions before moving onto the next section.

First, expose a material on our `SpatialMappingManager` class to allow us to easily update the material assigned to our surface meshes:

```
public Material SurfaceMaterial;
```

Now, add the following code to our `OnDataReady` method:

```
void OnDataReady(SurfaceData bakedData, bool outputWritten, float
elapsedBakeTimeSecond)
{
    if (!outputWritten)
        return;

    GameObject surface;
    if (cachedSurfaces.TryGetValue(bakedData.id.handle, out
    surface))
    {
        MeshRenderer renderer = surface.GetComponent<MeshRenderer>
        ();
        if (SurfaceMaterial != null)
        {
            renderer.sharedMaterial = SurfaceMaterial;
        }
    }
}
```

Here, we are simply assigning the referenced material to the surface renderer, when not null. Our final addition is giving us the ability to hide and show the surfaces at runtime. For this, we will add a property that will update all the cached surfaces, updating their visibility based on the set value. Let's do that now by adding the following code to our `SpatialMappingManager` class:

```
bool _surfacesVisible = true;

public bool SurfacesVisible
{
    get { return _surfacesVisible; }
    set
    {
        _surfacesVisible = true;

        foreach (KeyValuePair<int, GameObject> entry in cachedSurfaces)
        {
            MeshRenderer renderer =
entry.Value.GetComponent<MeshRenderer>();
            renderer.enabled = _surfacesVisible;
        }
    }
}
```

Finally, update the `OnDataReady` method to take account of these changes:

```
void OnDataReady(SurfaceData bakedData, bool outputWritten, float
elapsedBakeTimeSecond)
    {
        if (!outputWritten)
            return;

        GameObject surface;
        if (cachedSurfaces.TryGetValue(bakedData.id.handle, out
        surface))
        {
            MeshRenderer renderer = surface.GetComponent<MeshRenderer>();
            if (SurfaceMaterial != null)
            {
                renderer.sharedMaterial = SurfaceMaterial;
            }
        renderer.enabled = SurfacesVisible;
        }
    }
```

We have done a lot in this section--we explored how to scan and visualize our environment. In the next section, we will look at an alternative technique using preexisting components.

Spatial mapping components

In the preceding section, we learned how to scan and visualize the environment using the low-level API. In this section, we will do it all again, but this time using the prebuilt components.

We will continue using the same scene, but will disable our `SpatialMappingManager` script. Do this by selecting **SpatialMapping** in the **Hierarchy** panel and unchecking the attached **SpatialMappingManager (Script)**.

With **SpatialMapping** selected, add the **Spatial Mapping** component via the **Component | AR | Spatial Mapping Collider (Script)** menu.

Spatial Mapping Collider (Script) takes care of most code we wrote in the preceding section. It periodically manages polling for surface changes and manages the creation, updation, and deletion for each surface with respect to the change observed. When new surfaces are observed, the details of the surface are baked into the `MeshFilter` of the associated GameObjects and a `MeshCollider` is attached, essentially making the real world tangible in our virtual world. Update the **Spatial Mapping Collider** component properties with the following values:

- Set **Mesh Layer** to `SpatialSurface`; this assigns our `SpatialSurface` layer to the instantiated GameObject.
- Select and drag the **SpatialMapping** GameObject from the **Hierarchy** panel onto the **Surface Parent** field of the **Spatial Mapping Collider** component in the **Inspector** panel. Newly created GameObjects will be a child of this.
- Update **Time Between Updates** to 3; this is the duration the environment will be scanned for before reconstructing the surfaces.
- Set the **Bounding Volume Type** to **Sphere** and set **Radius In Meters** to **2**; similar to the preceding section, this describes the bounding volume for this instance of **Spatial Mapping**.

The following is a screenshot of the **SpatialMapping** component:

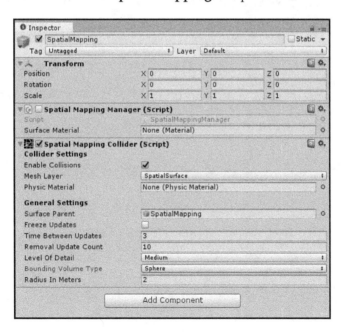

At this stage, we have an interactive environment, but it is invisible. Even though we are not required to visualize the surfaces in our application, visualizing is helpful in understanding how HoloLens sees the world and useful when debugging. So, let's go ahead and add the ability to visualize the surfaces.

Similar to the **Spatial Mapping Collider**, this is simply a matter of adding a component to our scene. With **SpatialMapping** selected, select the **Component** | **AR** | **Spatial Mapping Renderer** menu item. This will add the component directly to your **SpatialMapping** GameObject. On first glance, the properties look similar to those of the **Spatial Mapping Collider** component; that is because they are. **Spatial Mapping Collider** and **Spatial Mapping Renderer** work independently; you can use them together or separately. Let's go ahead and configure **Spatial Mapping Renderer** to match our **Spatial Mapping Collider** component; in the **Inspector** panel, update the properties with the following values:

- Set the **Custom Rendering Setting** to `Custom Material`; this will default to the **SpatialMappingWireframe**, the default material used in the previous section.
- Select and drag the **SpatialMapping** GameObject from the project's **Hierarchy** panel to the **Surface Parent** property. This is, again, the parent GameObject that surfaces will be nested under.
- Set the **Time between Updates** to 3.
- Finally, set the **Bounding Volume Type** to Sphere and **Radius In Meters** to 2.

Your **Inspector** panel for **SpatialMapping** should look similar to the following screenshot:

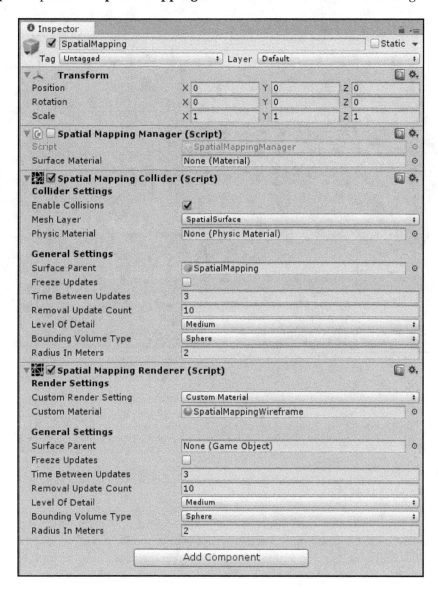

Well done, you managed to achieve everything we did in the last section without writing a single line of code. Now go ahead and build and deploy to see whether you can spot any difference between the two approaches:

With access to the physical world, our next task is to place the bin on the ground, which is the topic of the next section.

Essentially, the approach you use to scan the environment is up to you, but the following chapters will be using which approach for reference.

Placing our bin

Unlike Virtual Reality, where the experience is independent of your physical space, MR is concerned mainly with enhancing it; therefore holograms, to be effective, need to look, act, and behave like real objects. This involves precisely positioning and orienting holograms to places in the world that are meaningful to the user.

In this section, you will learn one approach for placing holograms in the real world, allowing the user to manually place the hologram onto a horizontal surface. In the next chapter, we will further explore placement by looking at another approach that tries to automatically place the hologram in the environment based on a set of specified constraints. Let's first start by discussing what we are trying achieve and a high-level description of the approach we will take.

Our goal is to simply place the bin somewhere on the floor that is appropriate for the user to throw paper into it. In this context, we define appropriate to mean placing the bin between a minimum and maximum distance from the user and somewhere the user has a clear line of sight, avoiding positions where the user and bin are obstructed by any surface.

We will have the user manually place the bin with our assistance; during placement, the bin will follow the user's gaze and, when in a valid position, allow the user to place the bin using the tap gesture. We consider a valid position to be one where there is an approximately flat surface underneath and the area is large enough for the bin to be placed onto.

Let's begin by first updating the **Layer** assigned to the **Bin** and applying the change to its prefab; we do this to have a convenient way of recognizing the bin when the user is gazing at it. Select the **Bin** from the project's **Hierarchy** panel in the **Inspector** locate the **Layer** dropdown and select the layer we created called `Hologram`. Select **Yes** when asked to apply to children. Now, click on the **Apply** button to apply those changes to the prefab. Finally, delete the bin from the scene by right-clicking on the **Bin** and selecting **Delete.**

Managing state

Before we can place the bin, we need to have a surface; furthermore, before we can throw paper at the bin, we need to have placed the bin. To capture these dependencies, we will create a script responsible for transitioning the application from one state to the next.

We will create a GameObject that will act as a container for all our scene-wide scripts. Create a new empty GameObject via the **GameObject | Create Empty** menu. Rename it to `Controllers` by right-clicking on the newly created GameObject in the project's **Hierarchy** panel, selecting **Rename**, and typing in `Controllers`. This will be the GameObject we attach our controller scripts to.

Now, let's create a script called `SceneController` that will be responsible for managing the states (scanning, placing, and playing) for this application. With the `Controllers` GameObject selected, click on the **Add Component** button that is visible on the **Inspector** panel and then select the **New Script** button and enter the name `SceneController`. Double-click on the `SceneController` script shown in the **Project** panel to open it in Visual Studio.

Within our `SceneController` class, create an enum for the all possible states of your game: scanning the environment, placing the bin, and playing the game:

```
public enum State
{
    Scanning,
    Placing,
    Playing
}
```

Add the associated property that will signal a method when updated:

```
State _currentState = State.Scanning;

public State CurrentState
{
    get { return _currentState; }
    set
    {
        _currentState = value;
        OnStateChanged();
    }
}

void OnStateChanged()
{

}
```

Our `OnStateChanged` method will be responsible for handling state changes. We need a way to transition from the state scanning to placing; in this context, we consider the scanning to be complete once we have observed enough of the environment to place the bin. There are many ways of quantifying enough; one approach might be determining whether you have enough surface area, or have found enough of a specific type of surface, such as vertical surfaces, but we will keep things simple in this instance and define enough as an elapsed amount of time after starting the scan.

To control the starting and stopping of scanning, we must first be able to communicate to the `SpatialMappingManager`, which is responsible for managing the scanning. There are many strategies for how components communicate between each other, but a common practice adopted by many Unity developers is to use the Singleton pattern. In the `SpatialMappingManager` class, add the following code:

```
public static SpatialMappingManager SharedInstance
{
    get
    {
        if(_sharedInstance == null)
        {
            _sharedInstance =
            GameObject.FindObjectOfType<SpatialMappingManager>();
        }

        if(_sharedInstance == null)
        {
            GameObject instanceGameObject = new
            GameObject(typeof(SpatialMappingManager));
            _sharedInstance =
            instanceGameObject.AddComponent<SpatialMappingManager>();
        }

        return _sharedInstance;
    }
}
```

The Singleton pattern is a design pattern that restricts to a single instance of a class. It is often useful when one object is responsible for coordinating actions across a system, and provides a convenient way of messaging between components.

With our static property in place, we now have access to the class `SpatialMappingManager` from within our `SceneController` class. Jump back into the **SceneController (Script)** and make the following amendments:

```
public enum State
{
    Scanning,
    Placing,
    Playing
}

public float scanningTime = 10f;

State _currentState = State.Scanning;
```

```
    public State CurrentState
    {
        get { return _currentState; }
        set
        {
            _currentState = value;
            OnStateChanged();
        }
    }

    void OnStateChanged()
    {
        StopAllCoroutines();

        switch (_currentState)
        {
            case State.Scanning:
                StartCoroutine(ScanningStateRoutine());
                break;
            case State.Placing:
                StartCoroutine(PlacingStateRoutine());
                break;
        }

    }

    void Start()
    {
        CurrentState = State.Scanning;
    }

    IEnumerator ScanningStateRoutine()
    {
        SpatialMappingManager.SharedInstance.IsObserving = true;
        yield return new WaitForSeconds(scanningTime);
        CurrentState = State.Placing;
    }

    IEnumerator PlacingStateRoutine()
    {
        SpatialMappingManager.SharedInstance.IsObserving = false;
        SpatialMappingManager.SharedInstance.SurfacesVisible = false;yield
return null;
    }
```

We first expose the `scanningTime` variable, making it available so that it can be easily tweaked from within the editor. This, as the name suggests, determines how long we reside in the scanning state.

Next, we create a switch statement within the `OnStateChanged` method to handle each state for the scanning and placing states, we spawn a coroutine to manage each independently. As with component communication (and programming in general) there are many alternatives; a similar approach is to embed conditional logic within the `Update` method. The approach taken is a personal preference and using coroutines allows the clean separation of state flows by encapsulating them in individual methods.

`ScanningStateRoutine` simply starts scanning by setting the `IsObserving` property of the `SpatialMappingManager` to `true`, and then waits for the specified time assigned to `scanningTime` before transitioning to the placing state.

The `PlacingStateRoutine` stops scanning by setting the `IsObserving` property of the `SpatialMappingManager` to false as well as hiding the surfaces by setting its `SurfaceVisible` property to false; next we will implement the components that will allow the user to place the bin, starting with gaze--the topic of the next section.

Implementing gaze

You can often infer someone's intent from what they're looking at, or in other words, their gaze. For example, people normally direct conversation by looking at the person they're conversing with (most of the time). Here, we are interested in where the user would like to place the bin; we will do this by attaching the bin to their gaze and follow it around, hugging the surface of the floor; when it is in a valid location, we will allow the user to place the bin by performing the tap gesture.

We will implement gaze using two components. The first will be responsible for inferring and updating the gaze based on the user's head position and orientation as well as assigning the gaze endpoint (where it hits the surface) to a publicly accessible property. The second component will be a visual representation of the gaze, the cursor, and will be used to provide feedback to the user to help them better understand what is currently in focus.

Create a script called `GazeController`; this script will be responsible for updating the user's gaze. With the `Controllers` GameObject selected, click on the **Add Component** button visible on the **Inspector** panel and then select **New Script** to create a new script. Enter the name `GazeController` and double-click to open it in Visual Studio. Add the following code:

```
public class GazeController : MonoBehaviour {

    private static GazeController _sharedInstance;

    public static GazeController SharedInstance
    {
        get
        {
            if (_sharedInstance == null)
            {
                _sharedInstance =
                GameObject.FindObjectOfType<GazeController>();
            }

            if (_sharedInstance == null)
            {
                GameObject instanceGameObject = new
                GameObject(typeof(GazeController).Name);
                _sharedInstance =
                instanceGameObject.AddComponent<GazeController>();
            }

            return _sharedInstance;
        }
    }
}
```

As we did previously, make the `GazeController` a singleton so that we can easily access it from other scripts. Now, add the following variables and properties to our `GazeController` class:

```
public float MaxDistance = 5f;

public LayerMask RaycastLayers = (1 << 31) | (1 << 30) | (1 << 5);

public Vector3 GazeHitPosition { get; private set; }

public Vector3 GazeHitNormal { get; private set; }

public Transform GazeHitTransform { get; private set; }
```

Also make the amendments as shown in the folowing code snippet:

```
public Vector3 GazeDirection
    {
        get
        {
            return Camera.main.transform.forward;
        }
    }

    public Vector3 GazeOrigin
    {
        get
        {
            return Camera.main.transform.position;
        }
    }
```

Here, we define a max distance that restricts how far the user can gaze, a mask to limit it to those objects we want the user to gaze at, and finally, some properties to capture the current gaze, including the exact point of the gaze (GazeHitPosition), the surfaces or the object's normal direction (GazeHitNormal), and the associated transform (GazeHitTransform), if one exists. GazeDirection and GazeOrigin are convenient properties giving us access to the user's current facing direction (GazeDirection) and position (GazeOrigin).

Our final code snippet for GazeController will be responsible for updating the gaze properties at each frame:

```
void Update () {

        RaycastHit hit;

        if (Physics.Raycast(GazeOrigin, GazeDirection, out hit,
        MaxDistance, RaycastLayers))
        {
            GazeHitPosition = hit.point;
            GazeHitNormal = hit.normal;
            GazeHitTransform = hit.transform;
        }
        else
        {
            GazeHitPosition = GazeOrigin+ (GazeDirection *
            MaxDistance);
            GazeHitNormal = GazeDirection;
            GazeHitTransform = null;
        }

    }
```

Here, we cast a ray using `Physics.Raycast`, passing the user's current position (GazeOrigin), and facing direction (GazeDirection), along with the maximum distance to cast a ray (MaxDistance) and layers (RaycastLayers). This call will return `true` if it collides with an object bundling the details in the `hit` parameter, in which case we update the gaze properties. When no collision is detected, we simply revert to some default values.

That's it for the `GazeController`; let's now move onto implementing the visual component of the gaze, the cursor. Like `GazeController`, the cursor will be a singleton and responsible for updating the position and orientation of the assigned `GameObject` (the visual representation of the cursor), with the position and orientation determined by the gaze properties set by `GazeController`.

You can think of our gaze and cursor synonymous with the mouse pointer, allowing the user to select and interact with virtual content. Like the mouse pointer, we want to provide sufficient feedback to let the user know what they can and cannot interact with. For our application, we will do this my changing the cursor's color, determined by whether the object the user is gazing at is interactive or not. Let's go ahead and implement the `Cursor`.

With the `Controllers` GameObject selected, click on the **Add Component** button from the **Inspector** panel and then select **New Script,** entering the name `CursorController`. Double-click on the `CursorController` script shown in the **Project** panel to open it in Visual Studio, and add the following code:

```
private static CursorController _sharedInstance;

public static CursorController SharedInstance
{
    get
    {
        if (_sharedInstance == null)
        {
            _sharedInstance =
GameObject.FindObjectOfType<CursorController>();
        }

        if(_sharedInstance == null)
        {
            GameObject instanceGameObject = new
GameObject(typeof(CursorController).Name);
            _sharedInstance =
instanceGameObject.AddComponent<CursorController>();
        }

        return _sharedInstance;
    }
```

```
    }
```

As we did earlier, we're using the Singleton pattern to provide a convenient way of getting access to the `CursorController`. Next, add the variables:

```
    public LayerMask InteractiveLayers = (1 << 30) | (1 << 5);

    public Color InteractiveColor = new Color(0.67f, 1f, 0.47f);

    public Color DefaultColor = new Color(1, 1, 1);

    public float HoverDistance = 0.05f;

    public GameObject Cursor;

Quaternion cursorRotationFix = Quaternion.AngleAxis(90, Vector3.right);
```

We want a way to differentiate between interactive and non-interactive elements. Here, we are setting a bitmask to include the `Hologram` (layer 31) and UI (5) layer (instead of layer names, we use their IDs), which we consider to be the interactive elements in our scene.

Next, we add two colors that we will transition to and from, depending on whether the user is gazing at an interactive or non-interactive object.

We add the `HoverDistance` variable, which determines how much padding to apply between the collision surface and cursor.

We then add the `Cursor` variable to hold a reference to the `Cursor` GameObject (the visual representation of the cursor) that we will be moving, and a rotation offset (`cursorRotationFix`) to orientate the cursor correctly onto surfaces.

Finally, add the following code to finish the `CursorController` class:

```
    void LateUpdate()
    {
        Cursor.transform.position =
        GazeController.SharedInstance.GazeHitPosition +
        GazeController.SharedInstance.GazeHitNormal * HoverDistance;
        Cursor.transform.up =
        GazeController.SharedInstance.GazeHitNormal;
        Cursor.transform.rotation *= cursorRotationFix;

        Color cursorTargetColor = DefaultColor;

        if (GazeController.SharedInstance.GazeHitTransform != null
        && (InteractiveLayers.value & (1 <<
    GazeController.SharedInstance.GazeHitTransform.gameObject.layer)) > 0)
```

```
    {
        cursorTargetColor = InteractiveColor;
    }

    Cursor.GetComponent<MeshRenderer>().material.color =
    Color.Lerp(
        Cursor.GetComponent<MeshRenderer>().material.color,
        cursorTargetColor,
        2f * Time.deltaTime);
}
```

We first update the position and orientation of the `Cursor` based on the `GazeManager` properties:

```
Cursor.transform.position = GazeController.SharedInstance.GazeHitPosition +
GazeController.SharedInstance.GazeHitNormal * HoverDistance;
Cursor.transform.up = GazeController.SharedInstance.GazeHitNormal;
Cursor.transform.rotation *= cursorRotationFix;
```

Then, we assign the relevant color based on whether the gaze is over an interactive element or not, and then interpolate toward the target color:

```
        Color cursorTargetColor = DefaultColor;

        if (GazeController.SharedInstance.GazeHitTransform != null
        && (InteractiveLayers.value & (1 <<
GazeController.SharedInstance.GazeHitTransform.gameObject.layer)) > 0)
        {
            cursorTargetColor = InteractiveColor;
        }

        Cursor.GetComponent<MeshRenderer>().material.color =
        Color.Lerp(
        Cursor.GetComponent<MeshRenderer>().material.color,
        cursorTargetColor,
        2f * Time.deltaTime);
```

Our last task to complete gaze is to assign an instance of the **Cursor** to the **Cursor** variable of `CursorController`. From the imported project's assets, find the **Cursor** prefab in the `Prefab` folder and drag it onto the **Scene**. Now, drag the **Cursor** GameObject from the **Hierarchy** panel to the `CursorController`'s **Cursor** variable.

Your final scene should look similar to the following screenshot:

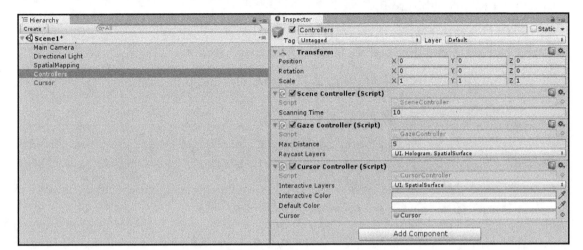

We've made some good progress. Go ahead and build and deploy to the emulator or device to see how far we have come; gaze around to get a feel for how the gaze and cursor components assist in helping you better understand what is currently in focus.

Finding a "good" place for the bin

As discussed earlier, our goal here is to provide a way for the user to manually place the bin; our approach is to attach a bin to the user's gaze (as implemented earlier) and follow it around, updating its position and orientation when in a valid position. Once happy, the user can place the bin using the tap gesture. In this section, we will implement this functionality; in the next section, we will implement the functionality used to detect the tap gesture to place the bin.

Before writing any code, let's quickly discuss how we will implement this. As the app transitions from scanning to placing, we will instantiate the Bin prefab and attach a component to it that will be responsible for making the bin follow the user's gaze. As we want our bin to obey the laws of physics, we will try to ensure that the bin is positioned on the ground rather than floating in the air or climbing up walls. To achieve this, we will do the following:

- Project the bin away from a wall when the user is gazing at a vertical surface
- Check that a valid surface is under the bin and, if so, update its position; valid here means a surface that exists and is reasonably flat

Let's start by creating the component responsible for following the gaze around and positioning the bin on the ground.

Click on the **Create** button on the **Project** panel and select **C# Script** to create a new script; name it `Placeable` and double-click to open it in Visual Studio.

Within the `Placeable` script, add the following variables and properties:

```
public LayerMask RaycastLayers = (1 << 31); // eq.
LayerMask.GetMask("SpatialSurface");

public float AngleThresholdToApplyOffset = 20f;

public float RaycastMaxDistance = 15f;

public float SurfaceAngleThreshold = 15f;

public float CornerAngleDifferenceThreshold = 10f;

public bool ValidPosition
{
    get;
    private set;
}

public bool Placed
{
    get;
    private set;
}
```

These variables describe the constraints we want the placement to adhere to; `RaycastLayers` determines the layer(s) the bin can rest on. `AngleThresholdToApplyOffset` is the angle threshold that determines whether we push the bin away from the wall or not. `SurfaceAngleThreshold` is the maximum angle of the surface the bin will rest on. `CornerAngleDifferenceThreshold` is used to determine whether the surface is flat or not, where each corner is compared against the other. Finally, we have the `ValidPosition` flag, indicating whether the bin is in a valid position as well as whether the bin has been placed.

Our final variables are used to animate the bin toward a target position and orientation; add the following to the `Placeable` script:

```
public float HoverDistance = 0.01f;

public float PlacementSpeed = 10f;

Vector3 targetPosition;

Vector3 targetNormal;

MeshFilter meshFilter;
```

`HoverDistance`, like our `Cursor` component, is an offset applied to the bin to avoid it hugging too tightly to a surface (avoiding possible Z-fighting); `targetPosition` and `targetNormal` are the current, valid, position, and orientation of our bin with the `PlacementSpeed` determining how quickly the bin transitions toward these values. Finally, `meshFilter` holds the reference to the GameObject's `MeshFilter`, a component that contains information about the mesh that we will use to get the bounds to determine where the corners are.

Add the following code in the `Start` method; this just gets a reference to the `MeshFilter` GameObject, and sets some default values for `targetPosition` and `targetNormal`:

```
void Start ()
{
    meshFilter = GetComponentInChildren<MeshFilter>();

    targetPosition = transform.position;
    targetNormal = transform.up;
}
```

For the rest, let's build it from the bottom up, starting with the `GetRaycastOrigins` method, a utility method that will return the center and corner positions of our bin, given a bounding box and offset:

```
Vector3[] GetRaycastOrigins(Bounds boundingBox, Vector3 originPosition)
{
    float minX = originPosition.x + boundingBox.center.x -
boundingBox.extents.x;
    float maxX = originPosition.x + boundingBox.center.x +
boundingBox.extents.x;
    float minY = originPosition.y + boundingBox.center.y -
boundingBox.extents.y;
    float maxY = originPosition.y + boundingBox.center.y +
```

```
boundingBox.extents.y;
        float minZ = originPosition.z + boundingBox.center.z -
boundingBox.extents.z;
        float maxZ = originPosition.z + boundingBox.center.z +
boundingBox.extents.z;

        return new Vector3[]
        {
            new Vector3(originPosition.x + boundingBox.center.x, minY,
originPosition.z + boundingBox.center.z), // centre
            new Vector3(minX, minY, minZ), // back left
            new Vector3(minX, minY, maxZ), // front left
            new Vector3(maxX, minY, maxZ), // front right
            new Vector3(maxX, minY, minZ) // back right
        };
    }
```

We use this mainly to determine whether the surface is even or not. With that in mind, let's implement the method responsible for determining whether the surface is flat. Add the following method to your `Placeable` script:

```
    bool IsSurfaceApproximatelyEven(Vector3[] raycastPositions)
    {
        float previousAngle = Vector3.Angle(raycastPositions[0],
raycastPositions[1]);

        for (int i=2; i<raycastPositions.Length; i++)
        {
            float angle = Vector3.Angle(raycastPositions[0],
raycastPositions[i]);
            if(Mathf.Abs(previousAngle - angle) >
CornerAngleDifferenceThreshold)
            {
                return false;
            }
            previousAngle = angle;
        }

        return true;
    }
```

This method iterates through all corners, calculating the angle for each with regard to the center, and returns `false` if any angle differences are larger than `CornerAngleDifferenceThreshold`.

The next method is the workhorse for this component; let's build this up piece by piece. Start by adding the method signature to the `Placeable` class:

```
bool GetValidatedTargetPositionAndNormalFromGaze(out Vector3 position,
out Vector3 normal)
{
    position = Vector3.zero;
    normal = Vector3.zero;

    return true;
}
```

The method is responsible for setting the position and normal values of the passed in parameters and returning a flag indicating whether a valid bin position resides where the user is currently gazing:

```
Bounds boundingBox = meshFilter.mesh.bounds;
Vector3 originPosition =
GazeController.SharedInstance.GazeHitPosition;
```

Next, we get the bounds of the mesh; these are in local space, so we need to use `originPosition` to transform them into world space:

```
if (Vector3.Angle(Vector3.up,
GazeController.SharedInstance.GazeHitNormal) >=
AngleThresholdToApplyOffset)
{
    originPosition += GazeController.SharedInstance.GazeHitNormal *
boundingBox.extents.z * 2.2f;
}

var raycastPositions = GetRaycastOrigins(boundingBox,
originPosition);
```

Before calling `GetRaycastOrigins`, we check to see whether the user is gazing at a vertical surface such as a wall; if so, we push the `originPosition` out from the wall to allow us to project down to the floor. This is done by comparing the angle between the normal surface the user is gazing at and the constant up vector with the `AngleThresholdToApplyOffset` threshold.

Next, we project a ray from the center of the bin to check whether a surface exists and to verify that the surface is reasonably flat; continue updating the `GetValidatedTargetPositionAndNormalFromGaze` method of `Placeable` with the following code:

```
RaycastHit hit;
```

```
        if (!Physics.Raycast(raycastPositions[0], -Vector3.up, out hit,
RaycastMaxDistance, RaycastLayers))
        {
            return false;
        }

        position = hit.point + (HoverDistance * hit.normal);
        normal = hit.normal;

        if(Vector3.Angle(Vector3.up, normal) > SurfaceAngleThreshold)
        {
            return false;
        }
```

If we collide with a surface, we set the position and normal parameters and verify that the surface is not too steep when tested against the `SurfaceAngleThreshold` variable we declared. If it is considered too steep, we return false.

Our final check is to see whether the surface is flat and if so, use our `IsSurfaceApproximatelyEven` method that we implemented. Add the following code to perform this final check:

```
        if (!IsSurfaceApproximatelyEven(raycastPositions)){
            return false;
        }
```

That's it for our `GetValidatedTargetPositionAndNormalFromGaze` method; the final piece of code is simply a question of calling this method and updating the target position and orientation when finding a valid position. Add the following code to wrap up this section:

```
    void Update () {
        Vector3 position;
        Vector3 normal;
        if(GetValidatedTargetPositionAndNormalFromGaze(out position,
        out normal))
        {
            ValidPosition = true;
            targetPosition = position;
            targetNormal = normal;
        }

        transform.position = Vector3.Lerp(transform.position,
        targetPosition, Time.deltaTime * PlacementSpeed);
        transform.up = Vector3.Lerp(transform.up, targetNormal,
        Time.deltaTime * PlacementSpeed);
    }
```

In each frame, we are checking for a new valid position and orientation; if they're found, we update our `targetPosition` and `targetNormal` variables and animate toward these values until updated.

We are almost there; check your progress by building and deploying to your device or emulator to see how the bin is dragged around by your gaze.

Where's our bin? You will have noted that our bin doesn't make an appearance in our scene. The reason for this is simply that we're not instantiating it when the state transitions from scanning to placing.

Jump back into the `SceneController` script and make the following amendments.

Add the `BinPrefab` instance variable; as the name suggests, this will hold the reference to the **Bin** prefab, which will be instantiated when we transition to the placing state, and the `Bin` property to hold the reference to the bin instance:

```
public GameObject BinPrefab;

public GameObject Bin
{
    get;
    private set;
}
```

Now, update the `PlacingStateRoutine` method with the following changes highlighted:

```
IEnumerator PlacingStateRoutine()
{
    SpatialMappingManager.SharedInstance.IsObserving = false;
    SpatialMappingManager.SharedInstance.SurfacesVisible = false;

    Bin = GameObject.Instantiate(
    BinPrefab,
    Camera.main.transform.position +
    (Camera.main.transform.forward * 2f) - (Vector3.up * 0.25f),
    Quaternion.identity);

    var placebale = Bin.AddComponent<Placeable>();

    while (!placebale.Placed)
    {
        yield return null;
    }

    GameObject.Destroy(placebale);
    Bin.AddComponent<WorldAnchor>();
```

```
        CurrentState = State.Playing;
    }
```

After transitioning into the placing state, we instantiate an instance of `BinPrefab` and assign it to our `Bin` property, positioning it slightly in front and below the user. We then add the `Placeable` component to the `Bin`, which takes care of placing the bin. Once placed, we remove the `Placeable` component and attach a `WorldAnchor` before transitioning to the playing state.

 To avoid drift and increase stability, world anchors can be attached to your stationary GameObjects. Once attached, HoloLens ensures that it maintains its position and rotation relative to the frame of reference. World anchors can also be persisted and shared with other devices, providing means that are efficiently built-- shared holographic spaces, something we will explore in later chapters.

With that set up, jump back into the editor to give our `SceneController` a reference to the **Bin** prefab. Select the `Controllers` GameObject from the **Hierarchy** panel, and then locate and drag the **Bin** prefab from the imported assets onto the `SceneController` **Bin** prefab variable in the **Inspector** panel.

Once again, build and deploy to your device or emulator to see your code in action; if everything is working, you should see something similar to the following:

Placing the bin

The final piece of functionality we need to place the bin is to allow the user to signal that they are happy with the bin's location, and afterward, to transition to the playing state. We will implement this by listening to the tap gesture and, once it's been detected--(with the bin in a valid position)--set the `Placeable` property `Placed` to `true` and ignore further updates.

If we consider the cursor to be synonymous with a mouse pointer, then gestures are synonymous with mouse buttons. Gestures provide a way for the user to input to the system, ideally as naturally and intuitively as possible. Microsoft is working on a catalog of standard interactions that they have made available in their library, these include the following:

- **Air-tap** (**press and release**): Used to select holograms (such as double-clicking on GUI systems)
- **Bloom**: A system gesture for bringing up the Start menu
- **Hold**: When pressed beyond a specified threshold, this can be used to pick up holograms
- **Manipulation**: Activated when pressed and dragged in an absolute movement; movement is normally 1:1, allowing for interactions such as resizing, rotating, and translating holograms
- **Navigation**: Like Manipulation, but differs in how it is used; where Manipulation is normally used to directly manipulate the hologram, Navigation is intended to provide a way of isolating on a single axis and is commonly used for User Interface interactions

All gestures follow a common pattern consisting of the following steps:

1. Creating a new `GestureRecognizer`.
2. Specifying the gestures you are interested in.
3. Registering to events for those gestures.
4. Beginning to capture gestures.

As discussed, we are interested in the (air-) tap gesture and one is detected, given that the bin is in a valid position, we flag the bin as being placed for the `SceneController` to handle.

Jump back into the `Placeable` script and, at the top, add the `UnityEngine.VR.WSA.Input` namespace to give us access to gesture APIs, and add the following instance variable:

```
GestureRecognizer tapGestureRecognizer;
```

Our first task is to specify the gestures we're interested in and start listening for gestures; we will add this to the `Start` method of the `Placeable` script. Make the following changes to your `Start` method:

```
    void Start ()
{
        meshFilter = GetComponentInChildren<MeshFilter>();

        targetPosition = transform.position;
        targetNormal = transform.up;

        tapGestureRecognizer = new GestureRecognizer();
tapGestureRecognizer.SetRecognizableGestures(GestureSettings.Tap);
        tapGestureRecognizer.TappedEvent +=
        TapGestureRecognizer_TappedEvent;
        tapGestureRecognizer.StartCapturingGestures();
}
```

In the preceding code snippet, we created our gesture recognizer, specified the events we're interested in, assigned a delegate for the specified event, and finally, start listening for the specified gestures. Now, let's implement the delegate that handles the event `TappedEvent`:

```
    void TapGestureRecognizer_TappedEvent(InteractionSourceKind source, int
    tapCount, Ray headRay)
    {
        if (ValidPosition)
        {
            Placed = true;
            tapGestureRecognizer.TappedEvent -=
            TapGestureRecognizer_TappedEvent;
        }
    }
```

The method in the preceding code snippet implements
the `GestureRecognizer.TappedEventDelegate` delegate, and the parameter gives you a
sense of how flexible the library is; for example, the `source` parameter indicates which
input triggered the event--This is of the `InteractionSourceKind` type and includes the
following values:

- Controller
- Voice
- Hand
- Other

When notified of a tap event, we first check to see whether the bin is in a valid position and,
if so, we set `Placed` to `true` and unregister our event listener from our gesture recognizer.
That's it!

Nothing but net - throwing paper

In the earlier sections, we were concerned with setting up the environment. Now, with our
environment scanned and the bin placed, we are ready to allow the user to throw paper
into, hopefully, the bin. As we did earlier, let's quickly outline our general approach before
writing any code.

One motivation for MR devices is the desire to eliminate the abstraction between computers
and the real world, including how we interact with them. We want to borrow as much from
the real world as possible to help deliver an intuitive and natural experience. In basketball,
the ball's velocity is determined by the angle and force of the player's throw; this is ideally
what we want to mimic.

HoloLens recognizes gesture input by tracking the position of both hands that are visible to
the device; hands are detected when they are either in the ready state (the back of the hand
with index finger up) or the pressed state (the back of the hand).
The field of view for gestures differs from the other cameras on the device in that it focuses
on the lower left and right regions in front of the device, known as the **gesture frame**. We
can access the states of the hands by registering for the relevant events of the
`InteractionManager` class; we will take advantage of this to implement our throw
mechanic. In this section, we will implement a class responsible for tracking the hands,
allowing the user to throw paper when we detect a finger being pressed and then released.
Velocity and direction will be derived from the change in distance of the tracked hands,
from when we detect a press to when we detect a release.

Let's start by creating the component responsible for tracking the user's hands and managing the throwing of the paper.

Select the `Controllers` GameObject in the **Hierarchy** pane, click on the **Add Component** button in the **Inspector** panel, select **New Script**, and name it `PlayerInputController`. Double-click on the `PlayerInputController` script to open it within Visual Studio.

With `PlayerInputController` open, start by adding the `UnityEngine.VR.WSA.Input` namespace.

Like the other controllers, make the `PlayerInputController` a singleton by adding the following code:

```
private static PlayerInputController _sharedInstance;

public static PlayerInputController SharedInstance
{
    get
    {
        if (_sharedInstance == null)
        {
            _sharedInstance =
            GameObject.FindObjectOfType<PlayerInputController>();
        }

        if (_sharedInstance == null)
        {
            GameObject instanceGameObject = new
            GameObject(typeof(PlayerInputController).Name);
            _sharedInstance =
            instanceGameObject.AddComponent<PlayerInputController>();
        }

        return _sharedInstance;
    }
}
```

Next, we want to register (and unregister) to the interaction events to allow us to track the hands' positions and state. Add the following code to your `PlayerInputController`:

```
void OnEnable()
{
    InteractionManager.SourceDetected +=
    InteractionManager_SourceDetected;
    InteractionManager.SourcePressed +=
    InteractionManager_SourcePressed;
    InteractionManager.SourceReleased +=
```

```
        InteractionManager_SourceReleased;
        InteractionManager.SourceUpdated +=
        InteractionManager_SourceUpdated;
        InteractionManager.SourceLost += InteractionManager_SourceLost;
    }

    void OnDisable()
    {
        InteractionManager.SourceDetected -=
        InteractionManager_SourceDetected;
        InteractionManager.SourcePressed -=
        InteractionManager_SourcePressed;
        InteractionManager.SourceReleased -=
        InteractionManager_SourceReleased;
        InteractionManager.SourceUpdated -=
        InteractionManager_SourceUpdated;
        InteractionManager.SourceLost -= InteractionManager_SourceLost;
    }
```

As discussed in the preceding code snippet, we want to be able to track the hands to calculate the force to apply to the paper. To track the state, we will implement a new class and a data structure to keep track of them. Add the following at the top, but within our `PlayerInputController` class:

```
    public class HandState
    {
        public uint id;
        public bool isPressed;
        public float pressedTimestamp;
        public Vector3 pressedPosition;
        public Vector3 currentPosition;
    }

    Dictionary<uint, HandState> trackedHands = new Dictionary<uint,
    HandState>();
```

This allows us to track the state of the user's hands; all we need to do now is to populate and update these objects from the events fired from `InterfactionManager`. To make our life easier, we will create two utility methods that will be responsible for adding and removing states from our `trackedHands` dictionary. Add the following methods to your `PlayerInputController` script:

```
    HandState GetHandState(InteractionSourceState state)
    {
        if(state.source.kind != InteractionSourceKind.Hand)
        {
            return null;
```

```
        }

        if (!trackedHands.ContainsKey(state.source.id))
        {
            trackedHands.Add(state.source.id, new HandState
            {
                id = state.source.id
            });
        }

        Vector3 handPosition;
        if (!state.properties.location.TryGetPosition(out
        handPosition))
        {
            RemoveHandState(state);
            return null;
        }

        HandState handState = trackedHands[state.source.id];
        handState.currentPosition = handPosition;

        return handState;
    }

    void RemoveHandState(InteractionSourceState state)
    {
        if (trackedHands.ContainsKey(state.source.id))
        {
            trackedHands.Remove(state.source.id);
        }
    }
```

The GetHandState is the most interesting of the two methods, it's responsible for creating a new state, if one doesn't already exist, and for updating the associate currentPosition of HandState. If it fails to obtain the hand's state, it will remove it from the trackedHands dictionary and return null, signifying an invalid state.

With our utility methods implemented, the InteractionManager event's methods are just used to update the hand's state:

```
    void InteractionManager_SourceDetected(InteractionSourceState
    state)
    {
        GetHandState(state);
    }

    void InteractionManager_SourcePressed(InteractionSourceState state)
```

```
    {
        HandState handState = GetHandState(state);

        if (handState == null)
        {
            return;
        }

        handState.pressedPosition = handState.currentPosition;
        handState.pressedTimestamp = Time.time;
        handState.isPressed = true;
    }

    void InteractionManager_SourceUpdated(InteractionSourceState state)
    {
        GetHandState(state);
    }

    void InteractionManager_SourceReleased(InteractionSourceState state)
    {
        HandState handState = GetHandState(state);

        if (handState == null || !handState.isPressed)
        {
            return;
        }

        handState.isPressed = false;
    }

    void InteractionManager_SourceLost(InteractionSourceState state)
    {
        RemoveHandState(state);
    }
```

Of interest are the `InteractionManager_SourcePressed` and `InteractionManager_SourceReleased` methods. Within the `InteractionManager_SourcePressed` method, we obtain the current state and update its `pressedTimestamp` and `pressedPosition`, and set `pressed` to `true`. These will be used in the `InteractionManager_SourceReleased` method to calculate the force used to throw our paper, which we will do now.

Head to the top of your `PlayerInputController` and add the following variables:

```
public GameObject PaperPrefab;

public float ThrowForceMultiplier = 2.0f;
```

Our GameObject `PaperPrefab` will hold reference to the **Paper** prefab we want to throw, which we will assign shortly. The `ThrowForceMultiplier` is a multiplier we will apply to the user's force, saving the user from having to do a workout when using our app. Within the `InteractionManager_SourceReleased` method of your `PlayerInputController` script, add the following code:

```
Vector3 distance = handState.currentPosition -
handState.pressedPosition;
float elapsedTime = Time.time - handState.pressedTimestamp;

Vector3 force = distance / elapsedTime;
force += Camera.main.transform.forward;
force *= ThrowForceMultiplier;

GameObject paper = GameObject.Instantiate(PaperPrefab);
paper.transform.position = handState.currentPosition;

paper.GetComponent<Rigidbody>().AddForce(force,
ForceMode.Force);
```

We initially determine the velocity (remember your physics class: velocity = distance/time) based on the change in the distance of the user's hand between when they performed the press and release gestures. We supplement this with the forward direction the user is facing and then apply our multiplier.
With our force now calculated, we are ready to project our holographic paper into MR. The next lines instantiate a new instance of our **Paper** prefab, set its position to the position of the throwing hand, and then add the force to the attached `Rigidbody`.

Our final task is to assign the **Paper** prefab to our `PlayerInputController` component and align this functionality with our app state.

Head back into the editor, select `SceneController` from the **Hierarchy** panel, locate the **CrumbledPaper** prefab from the **Project** panel, and drag it onto the `SceneController` **Paper** prefab variable.

As we only want this functionality available in the playing state, let's go ahead and disable it. Do this by clicking on the checkbox shown at the top of the `PlayerInputController` component attached to the `SceneController` GameObject.

We are almost there; our final task is to enable `PlayerInputController` when we transition into the playing state. Jump back into your **SceneController (Script)** and make the following amendments to the `OnStateChanged` method:

```
void OnStateChanged()
{
    StopAllCoroutines();

    switch (_currentState)
    {
        case State.Scanning:
            StartCoroutine(ScanningStateRoutine());
            break;
        case State.Placing:
            StartCoroutine(PlacingStateRoutine());
            break;
        case State.Playing:
PlayerInputController.SharedInstance.gameObject.SetActive(true);
            break;
    }

}
```

Here, we are simply activating `PlayerInputController` when the state transitions into playing. Now that it's complete, build and deploy to your device or emulator to see your hard work in action--time to shoot some hoops!

Summary

Congratulations for making it this far. We've covered a lot in this chapter, starting with looking at how to set up Unity for HoloLens development. We then stepped through building a complete game, looking at the major building blocks of a HoloLens app in Unity, including spatial mapping, gaze, and gestures.

In the next chapter, we will rebuild this project, but leveraging Microsoft's open source library, HoloToolkit-Unity, to accelerate development. We will also take our project further by looking at a technique for automatically placing holograms (our bin).

5
Building Paper Trash Ball Using Holotoolkit in Unity

In the previous chapter, we walked through building a holographic game for the HoloLens. Through this process, we gained experience using some of the core building blocks of MR applications, including using spatial mapping to understand the environment, making use of gaze and a cursor to infer the user's intent, placing holograms in the physical world, and user interaction using gestures. We also discussed the importance of adequate user feedback and its value in delivering an enjoyable user experience. In this chapter, we are going to do it all again, but this time, using HoloToolkit, an open source initiative from Microsoft to help accelerate HoloLens development. I hope that using the same concept will help illustrate how effective using the toolkit can be when developing for HoloLens and provide us with an opportunity to focus on improving the user experience. By the end of this chapter, you will know how to do the following:

- Use HoloToolkit to increase your productivity when developing HoloLens applications
- Automatically place your holograms in an environment based on a set of constraints

Adding HoloToolkit to your project

Our first job is to get HoloToolkit into our project. To do this, we will be loading a version of HoloToolkit-Unity (original source from `https://github.com/Microsoft/HoloToolkit-Unity`) that comes bundled with the project files for this book under the directory `HoloToolkit-Unity`.

The version of HoloToolkit used in this chapter was released in late 2016. You can clone this version directly from the original source by running the following commands in the Command Prompt:

git clone `https://github.com/Microsoft/HoloToolkit-Unity.git`

git checkout cd2fd2f9569552c37ab9defba108e7b7d9999b12

We now will create a Unity package so we can easily import it into our project. Once the HoloToolkit-Unity project is loaded in Unity, select everything in your **Project** panel. With everything selected, export it as a Unity package via **Assets** | **Export Package**. Leaving everything checked, click on the **Export** button.

The reason for selecting everything is to ensure we include the build and compilation option files `rsp`, `csc`, `gmcs`, and `smcs`.

With HoloToolkit now exported, it's time to create a new project and start making use of it. Assuming you still have Unity open, create a new project from the menu **File** | **New Project** and enter an appropriate name and location before clicking on the **Create project** button.

With your new project now open, import the HoloToolkit package by either double-clicking on the exported package or importing it via the menu **Assets** | **Import Package** | **Custom Package**, ensuring that you select the exported Unity package you created from the previous step. Leaving everything checked, click on the **Import** button to bring all the files into your project:

As we did before, we will import the project-related assets. From the menu, select **Assets** | **Import** | **Custom Packages**, navigate to where the book resources reside, and select the package `PaperTrashBallAssets_START.unitypackage`. Leaving everything checked, click on the **Import** button to import the assets into your package.

Credit and thanks to Daniele De Luca for the use of the models used in this project, available at `http://www.blendswap.com/blends/view/19629`.

With all the assets now added, let's begin by setting up Unity for developing for the HoloLens. HoloToolkit comes with some useful Editor utilities; we will use two now to help setup our environment. From the menu, select **HoloToolkit** | **Configure** | **Apply HoloLens Project Settings**. This will set your platform to **Windows 10 Universal**, set **Quality Settings** to **Fastest**, and enable **Virtual Reality Support**.

To perform spatial mapping, we require the spatial perception capability. Enable this via the **Capability Settings** dialog using the menu **HoloTookit** | **Configure** | **Apply HoloLens Capability Settings.** With the dialog open, check the **SpatialPerception** capability to enable it as in the following screenshot:

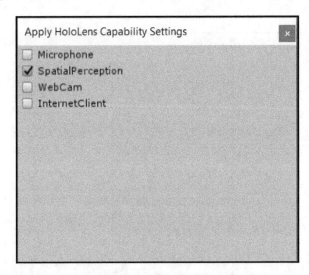

Now, let's set up the camera for HoloLens; you can do this using the menu **HoloToolkit** | **Configure** | **Apply HoloLens Scene Settings**. Selecting this will configure your current scenes **Main Camera** with the appropriate settings. Alternatively, you can use the prefab that comes with HoloToolkit. Because we already have **Main Camera**, we will use the menu option. From the menu, select **HoloToolkit** | **Configure** | **Apply HoloLens Scene Settings.**

Our final task is to create new layers that will be assigned to the scanned surfaces and holograms. From the menu, select **Edit** | **Project Settings** | **Tags and Layers**. This will open a panel, **Layers** as shown in the following screenshot, listing all the layers. As we did in the previous chapter, add SpatialSurface at index 31 and Hologram at index 30. SpatialSurface will be assigned to our surfaces and Hologram will be assigned to our holograms:

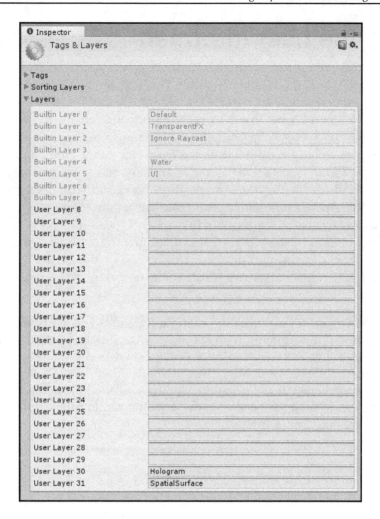

Now, with our environment set up, let's turn our attention to providing better feedback to the user.

No doubt we have all experienced applications where the feedback has been inadequate, leaving us feeling lost and, at times, frustrated. In our example, we will use the cursor as our main form of feedback, as we did before, but this time, providing some additional information, such as showing when we have detected the user's hand in the ready state. Along with this, we will also introduce a floating status text field that will tag along with the user's gaze, which will be used to keep the user informed of the current state of the application and could also include any other helpful information that might help the user progress smoothly through the application.

Keeping the user informed

Within the Editor, search for the `CursorWithFeedback` prefab in the **Project** panel's search bar. Once found, drag onto the **Hierarchy** panel. This almost takes care of the cursor. We just need to add two dependencies, namely, `GazeManager` and `HandManager`. Create a new GameObject via the menu **GameObject | Create Empty** and rename to `Gaze`; this will be the GameObject we attach these components to. With the `Gaze` GameObject selected, click on the **Add Component** button from within the **Inspector** panel, enter, and select the `GazeManager` script. The `GazeManager`, as the name implies, provides information about the user's gaze, including the direction and position. Next, still with the `Gaze` GameObject, selected, click on the **Add Component** button from within the **Inspector** panel, enter, and select the script `HandManager`. The `HandManager` script tracks the hands, and when in view broadcast the event `HandInView` along with maintaining the state, such as the number of hands currently in view, via public variables. This is used by the script `CursorFeedback` to update the visual state of the cursor to indicate to the user if their hand(s) have been detected or not.

HoloLens recognizes hands when they are either in the ready state (back of the hand facing you with the index finger up) or pressed state (back of hand facing you with the index finder down-as shown in the following figure). Otherwise, it will ignore them. Another consideration is the area the HoloLens tracks for hands. This area is called the **gesture frame**, an area which extends above, below, left, and right of the display frame where holograms appear.

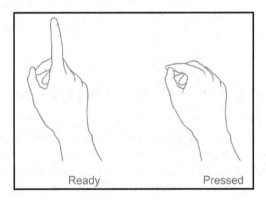

With just a few clicks of the mouse, we have managed to set up the cursor for our scene but we can take this further. Currently, the gaze and cursor mirror the absolute position of the user pose. This can require significant effort from the user when trying to select something small. We can improve this by slightly smoothing the movement, making it easier for the user to hold the cursor at a specific location. To do this, we will add a script that comes bundled with HoloToolkit. With the Gaze GameObject still selected, click on the **Add Component** button from within the **Inspector** panel, enter, and select the GazeStabilizer script. With this now added, our cursor, including the feedback of the state of the user's hands, is ready. Let's now move onto our floating status message.

The motivation for this message is to help keep the user informed of the current state of the application and provide any useful instructions to assist the user progressing through the experience. We will use the SimpleTagalong script available in HoloToolkit. The GameObjects attached with this script, such as our status message, will be dragged along with the user's gaze but, unlike the cursor, it is not rigid and will only update its position when at the edge of the user's field of view. This pattern ensures that the content is easily accessible for the user without obstructing their view. Our status message will consist of the parent GameObject and the nested GameObjects with a TextMesh component attached to display our messages.

Create a new GameObject via the menu **GameObject | CreateEmpty**; name this GameObject StatusText. With the StatusText GameObject selected, click on the **Add Component** button within the **Inspector** panel, enter, and select the script SimpleTagalong. The SimpleTagAlong script exposes a few properties that allow you to tweak its behavior, including:

- Tagalong Distance: This is the distance (in meters) in front of the camera the GameObject should be
- Enforce Distance: It enforces Tagalong Distance even if the GameObject didn't need to be moved, that is, it wasn't out of the view of the user
- Position Update Speed: How quickly the GameObject moves (meters/second)
- Smooth Motion: when checked, uses Smoothing Factor to interpolate the GameObject to a set target position.

We now will add the TextMesh component inside our StatusText GameObject. Right-click on StatusText inside the **Hierarchy** panel and select **Create Empty**. Rename the GameObject to Text by right-clicking on the GameObject and selecting the **Rename** menu item. With the Text GameObject selected, click on the **Add Component** button within the **Inspector** panel, enter, and select TextMesh. This will add TextMesh along with MeshRenderer to our Text GameObject, which will take care of rendering the text assigned to its Text property.

To have the text face the user, we will add the `Billboard` script. With the `Text` GameObject still selected, click on the **Add Component** button from within the **Inspector** panel, enter, and select `Billboard`.

We will now make some amendments to the `Text` GameObject properties. With it selected, make the following changes inside the **Inspector** panel:

- Set **Layer** to `Ignore Raycast` to avoid it interfering with the user's gaze
- In the **Transform** pane, set the **x** and **y Scale** to `0.005`
- In the **Text Mesh** pane, set **Anchor** to `Middle Center` and **Font Size** to `64`

To make our lives easier, we will create a script that acts as a singleton, giving us a convenient way of accessing and updating text from anywhere within our project. Select the `StatusText` GameObject in the **Hierarchy** panel and click on the **Add Component** button in the **Inspector** panel, select **New Script**, and enter the name `StatusText`. Double-click on the `StatusText` script from the **Project** panel to open it in Visual Studio. Once opened, make the following amendments:

```
using HoloToolkit.Unity;

public class StatusText : Singleton<StatusText>
{

    TextMesh textMesh;

    public string Text
    {
        get
        {
            return textMesh.text;
        }
        set
        {
            textMesh.text = value;
        }
    }

    void Start ()
    {
        textMesh = GetComponentInChildren<TextMesh>();
    }
}
```

As mentioned earlier, for convenience, we have made the class a singleton, so we can easily access it anywhere in our project. In the `Start` method, we search the children for the `TextMesh` component and, finally, wrap the `text` property of `TextMesh` in our own `Text` property. Your scene **Hierarchy** and **Project** panels should now resemble the following screenshot:

That concludes our brief tour on user feedback. Not extensive, but I hope this exercise has highlighted the importance of ensuring adequate feedback, especially for emerging platforms, such as the HoloLens. Let's continue our exploration of the HoloToolkit and see what tools are available for spatial mapping and understanding our environment.

Mapping our environment

In the previous chapter, we took two routes; the first using low-level APIs and the second simplifying the process by using the `SpatialMappingCollider` and `SpatialMappingRender` components. HoloToolkit simplifies this further. To add spatial mapping, we simply drag the `SpatialMapping` prefab onto the scene. Let's add this now; from the **Project** panel, enter `SpatialMapping` into the search bar. One of the results will be the prefab. Drag this one onto the **Hierarchy** panel to have it added to your scene, as shown:

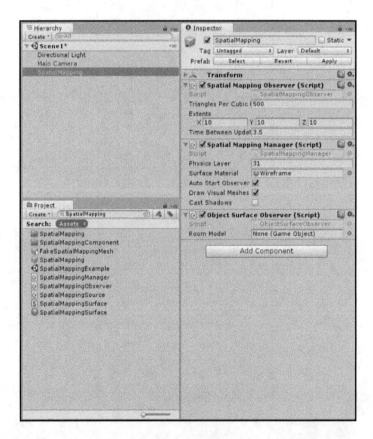

The properties should look familiar to you as they, are intentionally, similar to what we implemented in the previous chapter. **Spatial Mapping Observer** and **Spatial Mapping Manager** are responsible for both scanning and, optionally, making surfaces visible but unlike our implementation, it uses an axis aligned box for its bounding volume, and as opposed to sphere, this uses the property **Extends** to determine the volume. There is an additional component attached to the `SpatialMapping` prefab, **Object Surface Observer**. This component provides a convenient way to use previously baked environments from within the Editor.

Now is a good time to see your progress and view it on the device or emulator. Launch the **Build Settings** dialog by selecting the menu item **File | Build Settings.** Click on **Add Open Scenes** to include the current scene in the build and click on the **Build** button to build the project.

Once finished, open the root directory of the build and double-click on the solution (`.sln`) file to open Visual Studio and, finally, build and deploy to the device or emulator. Refer to `Chapter 9`, *Deploying Your Apps to the HoloLens Device and Emulator* for details.

As you can see, with minimal effort, we now have access to the physical world. Our next task is to try to better understand this physical world, so we can automatically place the bin. This is the focus of the next section.

Procedurally placing our bin

In the previous chapter, we had the user explicitly place the bin. This was achieved by having the bin follow the user's gaze, positioning itself on valid surfaces, and setting its position after detecting an air tap from the user. In this chapter, we will bypass the user completely and place the bin automatically.

To be able to place the bin, at a reasonable location, we first need to better understand the environment. Similar to how we interpret the world through objects rather than pixels, essentially we need a way to express the environment in a more meaningful way than the mesh data we receive from the `SurfaceObserver` to be able to effectively place the bin procedurally, for example, recognizing a surface as being the floor, where bins normally reside. Luckily, HoloToolkit provides us with the necessary tools to achieve this and we will spend the rest of this section, and a significant amount of this chapter, exploring two different approaches. The first will be using the utility class `SurfaceMeshesToPlanes`.

This utility converts surface mesh into a set of planes, tagging each with metadata defining the type of surface the plane belongs to, such as wall, floor, table, or ceiling. The second approach will be using the package SpatialUnderstanding, a helpful library that allows us to interrogate and search for places that adhere to a set of specified constraints.

SurfaceMeshesToPlanes

The core concept of this approach is converting the observed surfaces returned by SurfaceObserver into a set of planes using the script SurfaceMeshesToPlanes. The process also includes tagging the type of plane using some basic heuristics; for example, vertical planes are considered walls. The lowest horizontally plane is tagged as being the floor and the highest horizontal plane tagged as being the ceiling, an so on.

Let's start by creating a new empty GameObject, naming it SpatialProcessing, where we will attach all our spatial processing components to. Do this via the menu **GameObject | Create Empty**, and give it the name SpatialProcessing. As mentioned earlier, we will be making use of some of the tools bundled in with HoloTookit, namely, the SurfaceMeshesToPlanes and RemoveSurfaceVertices components. Details of what they do and how they are used will become clear as we walk through the implementation but, for now, just add these to our SpatialProcessing GameObject. Do this by selecting the SpatialProcessing GameObject in the **Hierarchy** panel and clicking on **Add Component** from within the **Inspector** panel. Add the components SurfaceMeshesToPlanes and RemoveSurfaceVertices to the SpatialProcessing GameObject by via the **Add Component** button from within the **Inspector** panel.

Let's create a new script to make use of the SurfaceMeshesToPlanes and RemoveSurfaceVertices components. With the SpatialProcessing GameObject selected, click on the **Add Component** button from within the **Inspector** panel and select the **New Script** option at the bottom of the drop-down; enter the name PlaneFinder. Your project should look similar to the following screenshot:

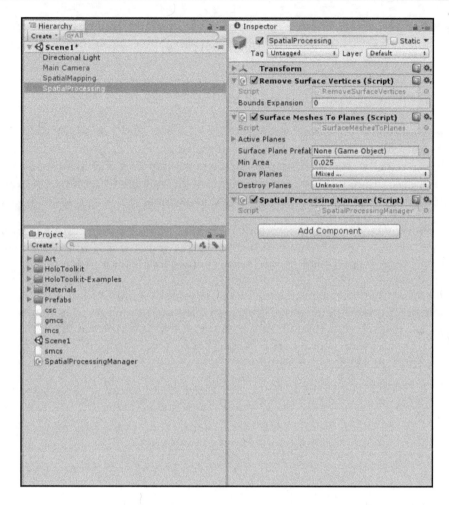

Before we jump into Visual Studio, let's briefly describe the properties exposed by the SurfaceMeshesToPlanes and RemoveSurfaceVertices components, starting with SurfaceMeshesToPlanes:

- Active Planes is a collection holding references to the currently generated planes.
- Surface Plane Prefab can be assigned a prefab that is substituted as the planes, providing us a way to style the planes.

- Min Area and Destroy Planes provide threshold to filter out unwanted planes. Destroy Planes is a bitwise mask that we can set to destroy any type of plane we are not interested in.
- Draw Planes, like Destroy Planes, is a bitwise mask such that we can specify what planes we want visible and hidden.

RemoveSurfaceVertices exposes one property, Bounds Expansion; to understand its purpose, it's helpful to know what this script does. RemoveSurfaceVertices implements functionality to remove vertices from the surface mesh that falls within a set of bounding areas passed in as parameters, usually, the planes with some padding. The Bounds Expansion property defines the padding added to each plane.

Let's move on to making this all work; double-click on PlaneFinder from within the **Project** panel to open it up in Visual Studio. At the top of the class, add the HoloToolKit.Unity namespace to gain access to the HoloToolKit classes.

As in the previous chapter, we need some way of measuring our level of understanding of the environment (from the perspective of the application). In the previous chapter, we simply used the elapsed scan time. In this example, we will use the notion of meeting some predefined criteria, where criteria here is defined as having found a plane (or planes) of a specific type and size. Let's begin fleshing out our PlaneFinder script by first defining a structure to encapsulate this criteria. Add the following code to the PlaneFinder class:

```
public class PlaneFinder : Singleton<PlaneFinder> {

    [Serializable]
    public struct PlaneCriteria
    {
        public PlaneTypes tag;
        public PlaneTypes planeType;
        public float minX;
        public float minY;
        public int minCount;
    }

    void Start () {
    }

    void Update () {
    }
}
```

The first notable feature of this snippet of code is that `PlaneFinder` inherits from `Singleton<PlaneFinder>`; `Singleton` acts as a generic singleton `MonoBehaviour`, providing us with a convenient way of making singletons, useful given that we made extensive use of them in the previous chapter. Next, we define the structure `PlaneCriteria` to describe our criteria. Adding the `Serializable` attribute to the structure makes it visible from within the Editor, allowing for a more convenient way of managing and tweaking our criteria from within the Editor. The field `tag` provides metadata, such as linking the criteria with some entity type in our application; for example, you could define criteria for a bin and another criterion for a score card. `PlaneType` describes the type of plane this criterion is concerned with; `minX` and `minY` defines the minimum size a plane can be; similarly, `minCount` defines the minimal planes that must be found before this criteria is satisfied. Next, we will define some utility methods within `PlaneCriteria` that will be used when evaluating planes discovered. First, we will implement a static method to calculate the area. Add the following code within the `PlaneCriteria` structure:

```
public static float CalculateArea(GameObject plane)
{
    return plane.transform.localScale.x *
plane.transform.localScale.y;
}
```

Here, we are just calculating the area of the surface area of the plane. Next, we will implement an evaluation method that will return true if the given collection of planes meets the defined criteria, otherwise, returns `false`. The evaluation method will delegate most of the work to another method responsible for filtering through a collection of planes, returning a subset that satisfies the criteria. Decomposing it like this means we can make use of the filtering logic in other places, such as finding suitable planes for our bin or any other hologram we may want to place. Add the following code to the `PlaneCriteria`:

```
public List<GameObject> GetPlanes(List<GameObject> planes)
{
    var localThis = this;

    var satisfactory = planes.Where(plane =>
    {
        return plane.GetComponent<SurfacePlane>() != null
            && (plane.GetComponent<SurfacePlane>().PlaneType &
localThis.planeType) > 0
            && plane.transform.localScale.x >= localThis.minX
            && plane.transform.localScale.y >= localThis.minY;
    });
    return satisfactory.ToList();
}
```

```
public bool Evaluate(List<GameObject> planes)
{
    return GetValidPlanes(planes).Count >= minCount;
}
```

The `Evaluate` method does nothing more than compare the size of the valid planes returned by the `GetPlanes` method with the `minCount` condition of the criteria. The `GetPlanes` searches over all planes and returns only those that meet the `planeType` and `minArea` conditions.

We can now filter planes; but currently, we have no planes to filter. Let's fix that now by making use of the `SurfaceMeshesToPlanes` class. As described earlier, this class processes the surface mesh observed from `SurfaceObserver` via `SpatialMappingManager`. The process is computationally expensive and, therefore, it is recommended to run after having stopped observing the environment and is consequently run on a separate thread with the results being returned to the designated delegate on the main thread. Our `PlaneFinder` class will be responsible for coordinating the activity of scanning the environment, processing the surfaces, and evaluating the results. If all the criteria has been met, then we will signal the process to be finished; otherwise, we return to scanning and repeat the process again. Let's convert these ideas into code starting with defining the variables and properties. From within our `PlaneFinder` class, add the following (just under the `PlaneCriteria` struct):

```
public PlaneCriteria[] spatialCriteria;

public float scanTime = 20f;

public bool Finished { get; private set; }

public Material scanningMaterial;

public Material defaultMaterial;

float scanningStratTime = 0f;

bool makingPlanes = false;
```

spatialCriteria defines our criteria and scanTime determines how long we observe the environment before passing the surfaces to SurfaceMeshestoPlanes to convert into planes before evaluating them based on spatialCriteria. Finished flags if we have successfully found planes that meet all our criteria; otherwise, we are still searching. scanningMaterial and defaultMaterial allow styling of the surface mesh during and after the process. scanningMaterial is assigned to the surface mesh when scanning while defaultMaterial thereafter. scanningStartTime is a timestamp of when scanning started and, finally, makingPlanes is a flag indicating if we are still waiting for SurfaceMeshestoPlanes to finish its task.

Next, we will assign a delegate to SurfaceMeshestoPlanes and begin scanning. Add the following to the Start method of our PlaneFinder class:

```
void Start () {
    SurfaceMeshesToPlanes.Instance.MakePlanesComplete +=
Instance_MakePlanesComplete;

    StartScanningEnvironment ();
}

void StartScanningEnvironment ()
{
    SpatialMappingManager.Instance.SurfaceMaterial = scanningMaterial;
    SpatialMappingManager.Instance.DrawVisualMeshes = scanningMaterial
!= null;

    scanningStratTime = Time.time;

    if (!SpatialMappingManager.Instance.IsObserverRunning())
    {
        SpatialMappingManager.Instance.StartObserver();
    }
}
```

The StartScanningEnvironment method is responsible for assigning the appropriate material and beginning the scanning process via SpatialMappingManager. Next, we will implement the MakePlanesComplete delegate method the delegate called by SurfaceMeshesToPlanes when it has finished processing all the surface mesh. Add the following code:

```
void Instance_MakePlanesComplete(object source, System.EventArgs args)
{
    makingPlanes = false;

    int satisfiedCriteriaCount = spatialCriteria.Where(criteria =>
```

```
criteria.Evaluate(SurfaceMeshesToPlanes.Instance.ActivePlanes)).Count();

    if (satisfiedCriteriaCount == spatialCriteria.Length)
    {
        Finished = true;

        SpatialMappingManager.Instance.SurfaceMaterial =
defaultMaterial;
        SpatialMappingManager.Instance.DrawVisualMeshes =
defaultMaterial != null;
    }
    else
    {
        StartScanningEnvironment();
    }
}
```

As mentioned earlier, this method is called once `SurfaceMeshestoPlanes` has finished processing. We evaluate each plane against each specified criterion via the statement:

```
int satisfiedCriteriaCount = spatialCriteria.Where(criteria =>
criteria.Evaluate(SurfaceMeshesToPlanes.Instance.ActivePlanes)).Count();
```

If the number of successful criteria evaluations are equal to the number of specified criteria, we know we have satisfied our requirements and flag the process as `Finished`; otherwise, we start the process all over again.

We also need a way to switch between scanning and processing. This will be done from within the `Update` method of our `PlaneFinder` class. Add the following code to the `Update` method:

```
void Update () {

    if (Finished)
    {
        return;
    }

    if(!makingPlanes && Time.time - scanningStratTime >= scanTime)
    {
        if (SpatialMappingManager.Instance.IsObserverRunning())
        {
            SpatialMappingManager.Instance.StopObserver();
        }

        makingPlanes = true;
        SurfaceMeshesToPlanes.Instance.MakePlanes();
```

```
    }
  }
```

Here, we are simply ignoring the update method if `Finished`; otherwise, we stop observing and begin processing if we are not already processing the surfaces and if the elapsed time from scanning is equal to or greater than `scanTime`.

Before finishing with our `PlaneFinder` class, let's add a method that will return a list of planes that satisfy the criteria associated with a given tag. This will allow us to pass in bin and have all the planes returned that our bin can be placed on. Add the following method to `PlaneFinder`:

```
public List<GameObject> GetPlanesForTag(string tag)
{
    var planes = new List<GameObject>();
    var selectedCriteria = spatialCriteria.Where(criteria =>
criteria.tag.Equals(tag, StringComparison.OrdinalIgnoreCase)).ToList();
    selectedCriteria.ForEach((critiera) =>
    {
planes.AddRange(critiera.GetPlanes(SurfaceMeshesToPlanes.Instance.ActivePla
nes).Where(plane => !planes.Contains(plane)));
    });

    return planes;
}
```

The method `GetPlanesForTag` takes a tag. From this, it finds all the associated criteria and then iterates through each criterion, evaluating each plane and returning all the planes that satisfied the criteria to the caller.

Now, with a collection of planes that satisfy our criteria, our next task is to place our hologram onto a plane. Like in the previous chapter, we will manage the scene state from within a `SceneController` script; back in the editor, create a new GameObject by clicking on the menu **GameObject** | **Create Empty** and name this object `Controllers`. This GameObject will be used to host all our controller scripts.

With the new GameObject selected, click on the **Add Component** button from within the **Inspector** panel, select **New Script**, and enter the name `SceneController`. Double-click on the `SceneController` script from within the **Project** panel to open it up in Visual Studio.

First, we will inherit from `Singleton` and then use an enumeration to encapsulate all the possible states that our application can be in. Make the following amendments to the `SceneController` script. Remember to add the namespace `HoloToolkit.Unity` and get access to the HoloToolkit classes:

```
public class SceneController : Singleton<SceneController> {

    public enum State
    {
        Scanning,
        Placing,
        Playing
    }

    State _currentState = State.Scanning;

    public State CurrentState
    {
        get { return _currentState; }
        set
        {
            _currentState = value;
            OnStateChanged();
        }
    }

    void OnStateChanged()
    {

    }

    void Start () {
        CurrentState = State.Scanning;
    }
}
```

This should look familiar to you. Assuming you have read the previous chapter, we first define the possible states for our application along with the property to store the current state. As we did in the previous chapter, we will handle the state changes in a separate method that, in turn, will start an associated coroutine. Let's implement what happens when we switch into the Scanning state. Add the following code:

```
void OnStateChanged()
{
    StopAllCoroutines();

    switch (_currentState)
    {
        case State.Scanning:
            StartCoroutine(ScanningStateRoutine());
            break;
        case State.Placing:
            break;
        case State.Playing:
            break;
    }

}

IEnumerator ScanningStateRoutine()
{
    StatusText.Instance.Text = "Look around to scan your play
    area";

    while (!PlaneFinder.Instance.Finished) yield return null;

    StatusText.Instance.Text = "";

    CurrentState = State.Placing;
}
```

From within our ScanningStateRoutine method, we first notify the user what is happening and then wait until PlaneFinder has satisfied the specified criteria by setting Finished to true. Once finished, we set CurrentState to the Placing state.

What happens in this state is what we will do next, but before moving on, let's see how effective `SurfaceMeshesToPlanes` is at converting surface meshes into planes. Jump back into the editor build and deploy to the simulator, emulator, or device to see how well it performs:

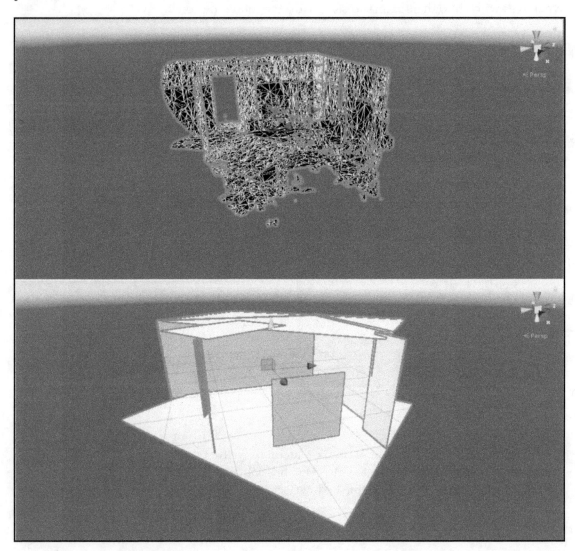

The preceding screenshots illustrates how effective `SurfaceMeshesToPlanes` is at dissecting surface mesh into a set of planes. It also highlights the fact that the planes are visible, not the look we were after. Let's quickly fix this and complete some final setup of our spatial processing before implementing the logic for placing the bin. From within the **Hierarchy** panel of the editor, select the `SpatialProcessing` GameObject, and within the **Surface Meshes to Planes** pane of the **Inspector** panel, set `Draw Planes` to `Nothing`. This will disable the `MeshRenderer` of each of the planes. Within the same **Inspector** panel, look for the **Spatial Processing Controller** pane and make the following changes:

1. Expand **Spatial Criteria** and add **1** to the **Size** field. This will increase our **Spatial Criteria** array to 1.
2. Enter the following details for the criteria.
3. a. Enter `bin` for the **Tag** Floor
4. b. Select `Floor` for **Plane Type**
5. c. Enter `0.32` for both **Min X** and **Min Y**

This defines the criteria for a plane suitable for our bin. In this example, we are simply placing one hologram but, as mentioned earlier, you could easily extend this to include multiple criteria for multiple holograms.

Finally, we will update the materials assigned to the surfaces for scanning and for when scanning completes. Click on the **Select** button (right-aligned circular button to this field) of **Scanning Material**. This will open a dialog showing a list of available materials. From this list, select the **Wireframe** material. Next, click on the **select** button next to **Default Material** and select the **Occlusion** material from the materials list. If you run the application again, you should see the wireframe of the surface mesh while the application is scanning that will disappear once the scan has finished and there are no visible planes. We now have everything to capture the environments surface. Convert them into a set of planes and return the suitable planes that satisfy our criteria. We next look at an approach for placing our bin onto one of these planes.

Placing holograms

Our next job is to place the hologram, the bin, onto a plane. But where? Before proposing a solution, let's first understand a little more of what we have available.

We currently have a set of planes that have satisfied our criteria. If you inspect a plane, you will notice that each plane faces away from the center; for example, the floor's forward direction points downwards. The following figure shows a ray cast at the center of each plane in the `-transform.forward` direction:

For our purposes, we want our hologram to be placed on top, which in this case, is the `-transform.forward` side of the plane, illustrated in the preceding screenshot. You will also notice that each plane uses local scale for its size. The following screenshot shows the **Inspector** panel of one plane:

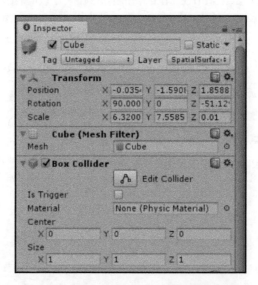

Now, with a better understanding of how the planes are orientated and sized, we will move onto proposing an approach for placing the hologram. Our goal here is to find a suitable place where our hologram(s) are not obstructed by any part of the environment, such as other planes or fragments of the surface, and in a convenient place for the user; for example, in a position that is not too close or too far and in user's view. We will use a brute-force search (also called an exhaustive search) to find suitable places through following the steps:

1. Define placement criteria for each hologram. This is how we will define and measure suitability.

2. Divide each surface of the plane, associated with the placement criteria, into a grid where each cell is considered a plausible position. The grid size will be relative to the size of the hologram being based.

3. Filter out positions that collide with the environment, such as the surface mesh or a plane.

4. Assign a cost to each position based on its distance and angle from the current user's facing direction.

5. Sort positions based on this cost.

6. Finally, select the position with the lowest cost and use this position to place the associated hologram (or positions and holograms if you are placing more than one hologram).

As we did with finding planes, we will encapsulate the responsibility of finding positions in a separate script. Back in the Editor, select the `SpatialProcessing` GameObject from the **Hierarchy** panel and click on **Add Component** from within the **Inspector** panel. Select **New Script** and enter the name `PlaceFinder` to create and attach a new script to the `SpatialProcessing` GameObject. Double-click on the `PlaceFinder` script from within the **Project** panel to open it in Visual Studio.

As we have done with the `PlaneFinder` script, add the namespace `Holotoolkit.Unity` and extend using the `Singleton` class to make it a singleton:

```
using HoloToolkit.Unity;

public class PlaceFinder : Singleton<PlaceFinder>
{

}
```

Next, we will define a structure that will describe our criteria for a placement with the variable tag to associate it with the plane from the previous state, prefab to hold reference to the associated hologram to be placed (our bin in this example), minDistance and maxDistance to define the ideal range from the user to place the hologram, and finally, count to describe how many holograms we want placed:

```
[Serializable]
public struct PlacementCriteria
{
    public string tag;
    public GameObject prefab;
    public int count;
    public float minDistance;
    public float maxDistance;
}
```

We will also add a property that will derive and return the bounding area based on the assigned prefab, which will be used when generating the grid when searching and positioning the hologram. Add the following property to the PlacementCriteria structure:

```
public Bounds Bounds
{
    get
    {
        Bounds? bounds = null;

        if(prefab.GetComponent<Collider>() != null)
        {
            bounds = prefab.GetComponent<Collider>().bounds;
        }
        else
        {
            Renderer[] renderers =
            prefab.GetComponentsInChildren<Renderer>();
            foreach (Renderer renderer in renderers)
            {
                if (bounds.HasValue)
                {
                    bounds.Value.Encapsulate(renderer.bounds);
                }
                else
                {
                    bounds = renderer.bounds;
                }
            }
        }
```

```
            return new Bounds(bounds.HasValue ? Vector3.zero :
            bounds.Value.center, bounds.HasValue ?
            bounds.Value.size : Vector3.zero);
    }
}
```

The bounds are determined by either the `Collider`, if attached, otherwise the bounding area encompassing all the GameObject's `renderers`. Next, we add an array to the `PlaceFinder` class to store these criteria:

```
public PlacementCriteria[] placmentCriteria;
```

Having `PlacementCriteria` decorated with the `Serializable` attribute makes it visible into the Editor. Let's jump back into the Editor and define our criteria. Back in the Editor, select the `SpatialProcessing` GameObject. Within the **Plane Finder** pane of the **Inspector** panel, increase the size of **Placement Criteria** to 1 and add the following values:

- Enter `bin` for the **Tag**
- Assign the `bin prefab` to the **Prefab** field
- Set **Count** to 1
- Set **Min Distance** to 1 and **Max Distance** to 2
- Finally, select `Hologram` and `SpatialSurface` for **Collision Layers**

With our placement criteria defined, let's jump back into Visual Studio and make use of it. Back in the `PlaceFinder` class, add the following structure, which will be used to store the places found during our search:

```
public struct Place
{
    public string tag;
    public Vector3 position;
    public Vector3 normal;
}
```

Each `Place` will represent a valid place for each `PlacementCriteria`. These will eventually be used to place the associated prefab. Let's add the variable to store these along with `CollisionLayers` used when checking for collisions at each position and the property `Finished` to flag when searching has finished:

```
public List<Place> foundPlaces = new List<Place>();

public LayerMask CollisionLayers = (1 << 31) | (1 << 30);

public bool Finished { get; private set; }
```

Searching will be triggered by `SceneController` through a public method. To keep the application responsive, we will wrap the search in a coroutine:

```
public void FindPlaces()
{
    Finished = false;

    StartCoroutine(SearchForPlaces());
}
```

The `SearchForPlaces` method will simply iterate over each `PlacementCriteria` and search each plane returned by the `GetPlanesForTag` method of the `PlaneFinder` class:

```
IEnumerator SearchForPlaces()
{
    foundPlaces.Clear();

    foreach (PlacementCriteria placementCriteria in
    placmentCriteria)
    {
        List<GameObject> planes =
        PlaneFinder.Instance.GetPlanesForTag(placementCriteria.tag);
         foreach (GameObject plane in planes)
         {
             FindPlacesOnPlane(placementCriteria, plane,
             foundPlaces);
         }

         yield return null;
    }

    Finished = true;
}
```

The bulk of the work is delegated to the method `FindPlaceOnPlane`, which follows the process described in the preceding section; first, dividing the plane into grid, where each cell is a candidate position, then removing all the positions that are occupied by other planes or surfaces in the environment, then assigning a cost to each position based on the defined criteria and finally, adding the number of required positions, as defined in the `count` variable of `PlacementCriteria` to the list. Once finished, we set `Finished` to true to flag the task as complete. Let's build this method up piece by piece, starting with getting all the positions:

```
void FindPlacesOnPlane(PlacementCriteria placementCriteria, GameObject
plane, List<Place> places)
    {
        var boundingRadius = Mathf.Max(placementCriteria.Bounds.size.x,
        placementCriteria.Bounds.size.y,
        placementCriteria.Bounds.size.z) * 0.5f;

        var allPositions = GetSurfacePositionsForPlane(plane,
        placementCriteria.Bounds, 0.5f);
        }
    }
```

We first get a bounding radius using the largest side of our `PlacementCriteria` bounds. This will be used later to represent the occupied space of our hologram and used to check for collisions for nearby planes and surfaces. We then pass the plane and bounds with an overlap to the `GetSurfacePositionsForPlane` method that will be responsible for dividing the plane into a grid and returning all the available positions. The overlap influences the spacing between each cell, passing in 1 will use the bounds of the hologram to determine the size of the cell (side by side), and passing in 1/2 will use half of the of the bounds of the hologram for the size of the stride, giving a denser grid and increasing the chances of finding a suitable place at the cost more computation. `GetSurfacePositionsForPlane` is one of the longer methods we have in this class but a significant amount of code simply calculates the offsets and size for each cell. Let's implement this method now. Add the following to your `PlaceFinder` class:

```
List<Vector3> GetSurfacePositionsForPlane(GameObject plane, Bounds
boundingBox, float overlap=1.0f)
    {
        Vector3 center = plane.GetComponent<BoxCollider>().center;
        Vector3 size = plane.GetComponent<BoxCollider>().size;

        float sx = -size.x;
        float ex = size.x;
        float sy = -size.y;
        float ey = size.y;
```

```
        float divX = plane.transform.localScale.x / (boundingBox.size.x
        * overlap);
        float divY = plane.transform.localScale.y / (boundingBox.size.z
        * overlap);

        float stepX = (size.x * 2f) / divX;
        float stepY = (size.y * 2f) / divY;
    }
```

Here, we simply determine divisions for the *x* and *y* axis along with the step size. To keep things relatively even, we need to account for the likely case where the plane is not divided into a whole number. We do this by adding margins based on the remainder, as shown:

```
        float countX = (size.x * 2f - (stepX * 2f)) / stepX;
        float countY = (size.y * 2f - (stepY * 2f)) / stepY;

        if(countX < 0 || countY < 0)
        {
            return new List<Vector3>();
        }

        float marginX = ((countX - (int)countX) / (int)countX) * 0.5f;
        float marginY = ((countY - (int)countY) / (int)countY) * 0.5f;
```

With our grid's starting and ending position, and cell size calculated, we now just need to iterate through each column and row to create an array of positions. Add the following code to our method GetSurfacePositionsForPlane:

```
    List<Vector3> surfacePositions = new List<Vector3>();

    for (float x = sx + stepX + marginX; x <= ex - stepX - marginX;
    x += stepX)
    {
        for (float y = sy + stepY + marginY; y <= ey - stepY -
        marginY; y += stepY)
        {
            surfacePositions.Add(plane.transform.TransformPoint(center +
            new Vector3(x, y, -size.z) * 0.5f) + -plane.transform.forward *
            0.001f);
        }
    }


        return surfacePositions;
```

We have plotting along the *x* and *y* axis, but our plane could be oriented in any direction. For this reason and to position relative to the plane's scale, we use the method `TrasnformPoint` to transform local points into world space. We also nudge the position slightly above the surface using the plane's `-transform.forward` direction, as discussed earlier, which positions the hologram just on top of the plane. That finishes our grid. The following screenshot gives us a bird's eye view of how the grid is currently laid out:

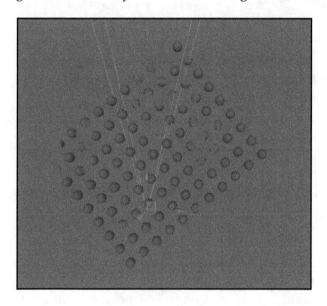

Our next job is to filter out positions that collide with existing planes and surfaces. Back in `FindPlacesOnPlane`, add the following state:

```
void FindPlacesOnPlane(PlacementCriteria placementCriteria, GameObject
plane, List<Place> places)
    {
        var boundingRadius = Mathf.Max(placementCriteria.Bounds.size.x,
        placementCriteria.Bounds.size.y,
        placementCriteria.Bounds.size.z) * 0.5f;

        var allPositions = GetSurfacePositionsForPlane(plane,
        placementCriteria.Bounds, 0.5f);
        var filteredPositions =
        FilterOutInvalidSurfacePositions(allPositions, -
        plane.transform.forward * boundingRadius, boundingRadius);
    }
```

We next pass all the positions, an offset, and the bounding radius to the method `FilterOutInvalidSurfacePositions`, expecting only the valid positions to be returned. The offset is used to offset each position slightly along the *y* axis, avoiding colliding with the surface and also providing a better representation of the position of our hologram. Add the following code to handle this validation:

```
List<Vector3> FilterOutInvalidSurfacePositions(List<Vector3>
surfacePositions, Vector3 positionOffset, float boundingRadius)
    {
        return surfacePositions.Where(surfacePosition =>
!CollidesWithEnvironment(surfacePosition + positionOffset, boundingRadius)
&& !IsObstructedFromUser(surfacePosition + positionOffset)).ToList();
    }
```

As described earlier, a valid position is one that does not collide with any other planes or surfaces and where nothing obstructs the user from this position. These checks are implemented in their own methods. Add the following method that will check for any collisions:

```
bool CollidesWithEnvironment(Vector3 position, float boundingRadius)
    {
        return Physics.OverlapSphere(position, boundingRadius,
        ~CollisionLayers, QueryTriggerInteraction.Ignore).Length > 0;
    }
```

We use one of Unity's collision detection methods against `CollisionLayers` defaulting to the `SpatialSurface` and `Holograms` layers. The method returns true if no collision has been detected.

Our next method checks to see whether anything is between the user and position. Add the following code to perform this check:

```
bool IsObstructedFromUser(Vector3 position)
    {
        Vector3 direction = (Camera.main.transform.position -
        position);
        float maxDistance = direction.magnitude;
        direction.Normalize();

        RaycastHit hitInfo;
        if(Physics.Raycast(position, direction, out hitInfo,
        maxDistance))
```

```
    {
        return hitInfo.transform != Camera.main.transform;
    }

    return false;
}
```

Here, we simply cast a ray toward the camera. If nothing obstructs the ray, then we return true, otherwise we return false. That now completes all that is necessary to validate a position. The following screenshot illustrates how our previous scene looks after validation has been applied, green signifying a valid position and red an invalid position:

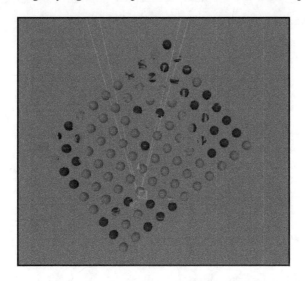

The next task requires us to assign a cost to each position, allowing us to pick the best (the one with the lowest cost) position. Cost, in this example, is based on the distance outside the specified minimum and maximum range to and from the user and also the angle away from the user's facing direction, assuming that where the user is looking is the most suitable and convenient direction to place our hologram. Back in our FindPlacesOnPlane method, add the following statement:

```
void FindPlacesOnPlane(PlacementCriteria placementCriteria, GameObject
plane, List<Place> places)
    {
        var boundingRadius = Mathf.Max(placementCriteria.Bounds.size.x,
        placementCriteria.Bounds.size.y,
        placementCriteria.Bounds.size.z) * 0.5f;

        var allPositions = GetSurfacePositionsForPlane(plane,
        placementCriteria.Bounds, 0.5f);
```

```
        var filteredPositions =
        FilterOutInvalidSurfacePositions(allPositions, -
        plane.transform.forward * boundingRadius, boundingRadius);
        var filteredAndSortedPositions =
        SortSurfacePosition(placementCriteria, filteredPositions);
    }
```

We now pass our filtered list of positions to `SortSurfacePositions` along with the associated `PlacementCriteria`, expecting a sorted array based on the passed in criteria, as described in the preceding code snippet. Copy in the following code to implement the `SortSurfacePositions` method:

```
List<Vector3> SortSurfacePosition(PlacementCriteria pc, List<Vector3>
surfacePositions)
    {
        var costs = surfacePositions.Select((position, index) =>
        {
            Vector3 cameraPos = new
            Vector3(Camera.main.transform.position.x, 0,
            Camera.main.transform.position.z);
            Vector3 pos = new Vector3(position.x, 0, position.z);

            Vector3 direction = (pos - cameraPos);
            float distance = direction.magnitude;
            direction.Normalize();

            float dot = Vector3.Dot(Camera.main.transform.forward,
            direction);
            float dotCost = Mathf.Pow(1f - dot, 2f);

            float minDistanceCost = 0;
            if(distance < pc.minDistance)
            {
                minDistanceCost = pc.minDistance/distance;
            }

            float maxDistanceCost = 0f;
            if(distance > pc.maxDistance)
            {
                maxDistanceCost = distance/pc.maxDistance;
            }

            float totalCost = dotCost + minDistanceCost +
            maxDistanceCost;

            return new { index = index, cost = totalCost };
        }).ToList();
```

```
costs.Sort((ca, cb) =>
{
    return ca.cost.CompareTo(cb.cost);
});

return costs.Select(cost =>
{
return surfacePositions[cost.index];
}).ToList();
}
```

The bulk of this method is taken up calculating a cost for each position. We perform this within a LINQ Select clause, returning a projection of the cost and index. Let's walk through each piece of the code to see how we are calculating the cost.

The following snippet finds the dot product between the user's facing direction and the direction of the position relative to the user's position. Two vectors facing the same direction will have a dot product of 1, perpendicular will return 0, and opposite directions will return -1. We subtract this dot product from 1 to give us a lower value, choosing the direction of the position relative to the user's facing direction, therefore, adding a higher cost for those that are out of view:

```
Vector3 cameraPos = new
Vector3(Camera.main.transform.position.x, 0,
Camera.main.transform.position.z);
Vector3 pos = new Vector3(position.x, 0, position.z);

Vector3 direction = (pos - cameraPos);
float distance = direction.magnitude;
direction.Normalize();

float dot = Vector3.Dot(Camera.main.transform.forward,
direction);
float dotCost = Mathf.Pow(1f - dot, 2f);
```

We next assign a cost to the distances that fall outside the minimum and maximum distances specified by `PlacementCritiera`:

```
float minDistanceCost = 0;
if(distance < pc.minDistance)
{
    minDistanceCost = pc.minDistance/distance;
}

float maxDistanceCost = 0f;
if(distance > pc.maxDistance)
{
```

```
            maxDistanceCost = distance/pc.maxDistance;
      }
```

Then, we sum the cost for each criteria before returning it along with the index associated with the current position:

```
    float totalCost = dotCost + minDistanceCost + maxDistanceCost;
    return new { index = index, cost = totalCost };
```

The last part of the method just sorts based on the calculated cost and, finally, rebuilds the list using the new ordering:

```
        costs.Sort((ca, cb) =>
        {
            return ca.cost.CompareTo(cb.cost);
        });

        return costs.Select(cost =>
        {
            return surfacePositions[cost.index];
        }).ToList();
```

It's worth noting that the preceding section is a soft constraint in contrast to the previous method, where we removed any positions that intersected with a surface or plane (hard constraint); here, we are just giving higher priority to the positions that satisfy these criteria rather than removing them. The following screenshot illustrates the sorting, increasing the opacity for those with higher priority:

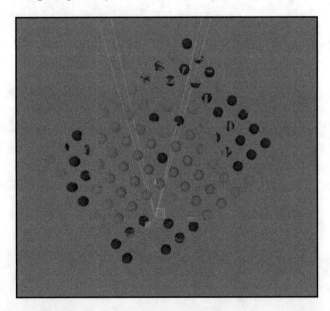

To finish off our `FindPlacesOnPlane` method, we simply need to create the `Place` objects for the number of positions as specified in the `count` variable of `PlaceCriteria`. Make the following amendments to finish off the `FindPlacesOnPlane` method:

```
void FindPlacesOnPlane(PlacementCriteria placementCriteria, GameObject
plane, List<Place> places)
    {
        var boundingRadius = Mathf.Max(placementCriteria.Bounds.size.x,
        placementCriteria.Bounds.size.y,
        placementCriteria.Bounds.size.z) * 0.5f;

        var allPositions = GetSurfacePositionsForPlane(plane,
        placementCriteria.Bounds, 0.5f);
        var filteredPositions =
        FilterOutInvalidSurfacePositions(allPositions, -
        plane.transform.forward * boundingRadius, boundingRadius);
        var filteredAndSortedPositions =
        SortSurfacePosition(placementCriteria, filteredPositions);

        for (int i = 0; i < Mathf.Min(placementCriteria.count,
        filteredAndSortedPositions.Count); i++)
        {
            places.Add(new Place
            {
                tag = placementCriteria.tag,
                position = filteredAndSortedPositions[i],
                normal = -plane.transform.forward
            });
        }
    }
```

The final method we will add to our `PlaceFinder` class will be used to place the holograms; here, we are simply iterating through each `placementCriteria`, finding the associated places, and instantiating the assigned prefab. Add the following code to finish off the `PlaceFinder` class:

```
public void PlaceGameObjects(GameObject container = null)
    {
        foreach (Place place in foundPlaces)
        {
            PlacementCriteria placmentCriteria =
            this.placmentCriteria.Where(pc =>
            pc.tag.Equals(place.tag)).First();

            GameObject go = Instantiate(placmentCriteria.prefab);
            go.transform.position = place.position;
            if (container != null)
```

```
        {
            go.transform.parent = container.transform;
        }
    }
}
```

Our last job is to hook this up to `SceneController`. Open the `SceneController` script in Visual Studio and make the following amendments to the `OnStateChanged` method:

```
void OnStateChanged()
{
    StopAllCoroutines();

    switch (_currentState)
    {
        case State.Scanning:
            StartCoroutine(ScanningStateRoutine());
            break;
        case State.Placing:
            StartCoroutine(PlacingStateRoutine());
            break;
        case State.Playing:
            break;
    }

}
```

Keeping our approach consistent, we start a coroutine to manage the `Placing` state. When entering this state, we want to request `PlaceFinder` to start searching for suitable places and wait until this task has been finished. Once finished, we place the items before transitioning into the `Playing` state. Let's put this down in code by adding the following method:

```
IEnumerator PlacingStateRoutine()
{
    PlaceFinder.Instance.FindPlaces();

    while (!PlaceFinder.Instance.Finished) yield return null;

    PlaceFinder.Instance.PlaceGameObjects();

    CurrentState = State.Playing;
}
```

With this method now implemented, our task for searching and placing the bin is complete. We're in a good place to build and test. The following screenshot shows the place for the default room. The user is shown as the red sphere:

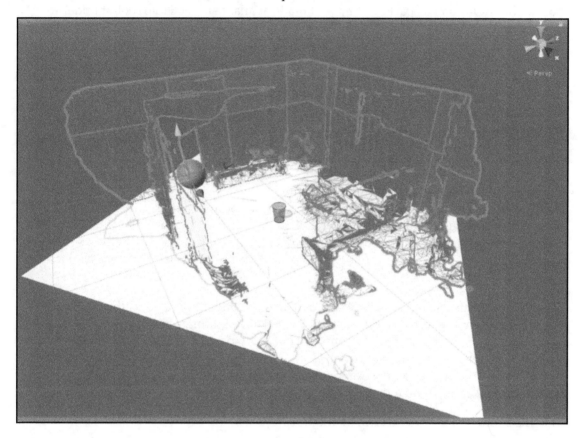

Spatial understanding

In the previous section, we walked through one approach of placing a hologram; the approach consisted of breaking the surface mesh into a set of planes, identifying an appropriate plane, then scoring each position on this plane before using the position with the best score. As you would expect, understanding and incorporating the physical world into MR applications is a common task. Fortunately, this task has been solved for us and made available in HoloToolkit. In this section, we will walk through how to make use of it in our application.

Asobo Studios faced the problem of better understanding the physical environment and finding suitable places for holograms when developing Young Conker, a HoloLens platform game where holographic character interacts and reacts naturally with the real world. Their solution has packaged up and made available in the HoloToolkit under the namespace `HoloToolkit.SpatialUnderstanding`. It comprises three primary modules:

- Topology for simple surface and spatial queries
- Shape for object detection
- Object placement solver for constraint-based placement

In this section, we will be mainly concerned with the object placement solver but I encourage you to experiment with the other modules using the example bundled with HoloTookit repository for reference.

The flow will be similar to the flow implemented in the previous section. We will scan, process, and search for suitable places based on some specified criteria. Instead of planes, we will be inspecting custom mesh generated by the `SpatialUnderstanding` library and once we have sufficiently scanned the environment, we will prompt the user to perform an air tap gesture before using `LevelSolver`, a utility that will search for places given a set of constraints, from `SpatialUnderstanding`.

Let's start by disabling and commenting out all the dependent parts from the previous section that are not required in this section. In the Unity Editor, select the `SpatialProcessing` GameObject from the **Hierarchy** panel and disable by clicking on the top left checkbox in the **Inspector** panel. Now, to disable the associated code, double-click on the `SceneController` script from within the **Project** panel to open it up in Visual Studio and make the following changes.

Note the `ScanningStateRoutine` method:

```
IEnumerator ScanningStateRoutine()
{
    StatusText.Instance.Text = "Look around to scan your play area";

    while (!PlaneFinder.Instance.Finished) yield return null;

    StatusText.Instance.Text = "";

    CurrentState = State.Placing;
}
```

Update it to the following:

```
IEnumerator ScanningStateRoutine()
{
    PlaySpaceScanner.Instance.Scan();

    while (PlaySpaceScanner.Instance.CurrentState !=
    PlaySpaceScanner.State.Finished)
    {
        if(PlaySpaceScanner.Instance.CurrentState ==
        PlaySpaceScanner.State.Scanning)
        {
            StatusText.Instance.Text = "Look around to scan your
            play area";
        }
        else if (PlaySpaceScanner.Instance.CurrentState ==
        PlaySpaceScanner.State.ReadyToFinish)
        {
            StatusText.Instance.Text = "Air tap when ready";
        }
        else if (PlaySpaceScanner.Instance.CurrentState ==
        PlaySpaceScanner.State.Finalizing)
        {
            StatusText.Instance.Text = "Finalizing scan (please
            wait)";
        }

        yield return null;
    }

    StatusText.Instance.Text = "";

    CurrentState = State.Placing;
}
```

This is functionally similar but with more states and delegating the task of finding suitable positions to PlaySpaceScanner. Now, replace the method PlacingStateRoutine with the following:

```
IEnumerator PlacingStateRoutine()
{
    PlaySpacePlaceFinder.Instance.FindPlaces();

    while (!PlaySpacePlaceFinder.Instance.Finished) yield return
    null;

    PlaySpacePlaceFinder.Instance.PlaceGameObjects();
```

```
        CurrentState = State.Playing;
    }
```

Like the previous method, we are retaining the functional flow but delegating the task to a new class.

Let's now address the errors by creating and stubbing out the classes `PlaySpaceScanner` and `PlaySpacePlaceFinder`. In the Editor, click on the **Create** button within the **Project** panel and select **C# Script**; enter the name `PlaySpaceScanner`. Again, click on the **Create** button, select **C# Script**, and enter the name `PlaySpacePlaceFinder`. Double-click on `PlaySpaceScanner` to open it up in Visual Studio. For now, we will just stub out each class method to remove compile time errors and will return to this shortly to fill in the details. With that in mind, add the following code to the `PlaySpaceScanner`, remembering to add the namespace `HoloToolkit.Unity`:

```
public class PlaySpaceScanner : Singleton<PlaySpaceScanner> {

    public enum State
    {
        Undefined,
        Scanning,
        ReadyToFinish,
        Finalizing,
        Finished
    }

    public State CurrentState
    {
        get;
        private set;
    }

    public void Scan()
    {

    }
}
```

Next, open the `PlaySpaceScanner` script and add the following code and the namespace `HoloToolkit.Unity`:

```
public class PlaySpacePlaceFinder : Singleton<PlaySpacePlaceFinder> {

    public bool Finished { get; private set; }

    public void FindPlaces()
    {

    }

    public void PlaceGameObjects()
    {

    }
}
```

With our classes now stubbed out, let's return to the Editor to prepare our scene to use `SpatialUnderstanding`.

As we did in the previous section, we will create an empty GameObject to host all the components necessary for `SpatialUnderstanding`. Let's create that now. Click on the **Create** button from the **Hierarchy** panel, select **Create Empty**, and enter the name `SpatialUnderstanding`. With our newly created GameObject selected, click on the **Add Component** button from within the **Inspector** panel, and enter and select `Spatial Understanding`. This will also include its dependencies `SpatialUnderstandingCustomMesh` and `SpatialUnderstandingSourceMesh`. Let's now briefly inspect each component, starting with `SpatialUnderstandingCustomMesh`.

`SpatialUnderstandingCustomMesh` is responsible for pulling through and reconstructing the retopologized surface mesh from the underlying library.

 The scanning process is an important step in the flow, as the higher level functions in the `SpatialUnderstanding` library depend on having the surfaces flat and walls at right angles. The `SpatialUnderstanding` library stores the space as a grid of 8cm sized voxel cubes. The mesh is extracted approximately every second using the isosurface from the voxel volume.

Setting the property `DrawProcessedMesh` to true will render the mesh using the assigned material to **Mesh Material**, talking of which, click on the **Object Selector** (circle button) to open the **Material** dialog, enter, and select `SpatialUnderstandingSurface`.

The next component, `SpatialUnderstandingSourceMesh`, is tasked with passing the observed surface mesh from `SpatialMappingObserver` to the `SpatialUnderstanding` library, which in turn processes the mesh.

And finally, the `SpatialUnderstanding` script is responsible for managing the scanning process. While we are here, uncheck **Auto Begin Scanning**, as we will manage the starting and stopping of the scanning based on the application state.

Let's now add our components `PlaySpaceScanner` and `PlaySpacePlaceFinder` to the `SpatialUnderstanding` GameObject. With the `SpatialUnderstanding` GameObject selected in the **Hierarchy** panel, click on the **Add Component** button from within the **Inspector** panel, and enter and select `PlaySpaceScanner`, and again for `PlaySpacePlaceFinder`. We will now return to the code to flesh out our `PlaySpaceScanner` script. Double-click on the `PlaySpaceScanner` script in the **Project** panel to open it in Visual Studio.

As the name suggest, our `PlaySpaceScanner` class will be responsible for scanning the environment, triggered to start scanning via `SceneController` when entering the `Scanning` state. It will continue scanning until we have sufficiently scanned enough-enough here defined as having scanned a surface that meets the specified minimum area threshold. Once the criteria have been meet, we will wait for the user to perform an air tap gesture before finalizing the scan and setting `Finished` to `true`, signaling to the `SceneController` to proceed to the next state. Add the following variables and properties to the `PlaySpaceScanner` class:

```
public float minAreaForComplete = 50f;

public float minHorizontalAreaForComplete = 10f;

public float minVerticalAreaForComplete = 10f;

private SpatialMappingObserver _mappingObserver;

public SpatialMappingObserver MappingObserver
{
    get
    {
        if(_mappingObserver == null)
        {
```

```
                _mappingObserver =
                FindObjectOfType<SpatialMappingObserver>();
            }

            return _mappingObserver;
        }
    }

    private SpatialUnderstandingCustomMesh
    _spatialUnderstandingCustomMesh;

    public SpatialUnderstandingCustomMesh
    SpatialUnderstandingCustomMesh
    {
        get
        {
            if (_spatialUnderstandingCustomMesh == null)
            {
                _spatialUnderstandingCustomMesh =
                FindObjectOfType<SpatialUnderstandingCustomMesh>();
            }

            return _spatialUnderstandingCustomMesh;
        }
    }
```

Our criteria, which determines whether we have sufficiently scanned environment, includes minimum area (`minAreaForComplete`) and minimum horizontal and vertical areas (`minHorizontalAreaForComplete`, `minVerticallAreaForComplete`). `MappingObserver` and `SpatialUnderstandingCustomMesh` are convenient getters that we will use within this class to get reference to the respective class instance. Next, add the following code for the `Start` method:

```
    void Start()
    {
        MappingObserver.SetObserverOrigin(Camera.main.transform.position);
        SpatialUnderstanding.Instance.ScanStateChanged +=
    Instance_ScanStateChanged;
    }
```

Here, we set the observer's origin to the camera's current position. This is used when setting the observer's bounding volume before each surface observation update, something we discussed in the previous chapter. We next register to state changes for the `SpatialUnderstanding` class, which are used to progress the experience forward. Let's define this delegate now. Add the following delegate to your `PlaySpaceScanner` class:

```
void Instance_ScanStateChanged()
{
    switch (SpatialUnderstanding.Instance.ScanState)
    {
        case SpatialUnderstanding.ScanStates.Scanning:
            SpatialUnderstandingCustomMesh.DrawProcessedMesh =
            true;
            CurrentState = State.Scanning;
            break;
        case SpatialUnderstanding.ScanStates.Finishing:
            CurrentState = State.Finalizing;
            break;
        case SpatialUnderstanding.ScanStates.Done:
            SpatialUnderstandingCustomMesh.DrawProcessedMesh =
            false;
            CurrentState = State.Finished;
            break;
    }
}
```

Most of what we are doing here is simply mapping the current state of `SpatialUnderstanding` to an equivalent state defined in the `PlaySpaceScanner` class with the addition of flagging whether to show or hide the processed mesh.

Before we move on, there is a slight issue with our current approach of showing and hiding the processed surface mesh. To make a hologram feel like it is part of the environment is to have it behave like real objects. An example of this is having it being occluded by real world objects when obstructing the user's view. Our current approach doesn't achieve this. We essentially remove the virtual representation of the physical world by hiding the surface mesh. We should rather update how it is rendered so it can occlude virtual objects that appear behind it (relative to the user). Let's fix this now. Head back to the top of the `PlaySpaceScanner` class and add the following code:

```
public Material defaultSurfaceMaterial;

public Material scanningSurfaceMaterial;

private Material _surfaceMaterial;

public Material SurfaceMaterial
```

```
        {
            get
            {
                return _surfaceMaterial;
            }
            set
            {
                _surfaceMaterial = value;

                SpatialUnderstandingCustomMesh.MeshMaterial =
                _surfaceMaterial;
                foreach(var surfaceObject in
                SpatialUnderstandingCustomMesh.SurfaceObjects)
                {
                    surfaceObject.Renderer.material = _surfaceMaterial;
                }
            }
        }
```

Here, we declare two variables to hold references for the material to be assigned to the processed surface mesh, one assigned when scanning and the other used as default. We also define a property that will hold the current mesh and propagate changes to all the surface renderers when set. Next, return to the Start method and make the following changes:

```
void Start()
{
        MappingObserver.SetObserverOrigin(Camera.main.transform.position);
SpatialUnderstanding.Instance.ScanStateChanged +=
Instance_ScanStateChanged;

SpatialUnderstandingCustomMesh.DrawProcessedMesh = true;
SurfaceMaterial = defaultSurfaceMaterial;
}
```

Within the Start method, we explicitly set the DrawProcessedMesh of SpatialUnderstandingCustomMesh to true and set the default surface material. Now, update the Instance_ScanStateChanged method with the following changes:

```
        void Instance_ScanStateChanged()
        {
            switch (SpatialUnderstanding.Instance.ScanState)
            {
                case SpatialUnderstanding.ScanStates.Scanning:
                    SurfaceMaterial = scanningSurfaceMaterial;
                    CurrentState = State.Scanning;
                    break;
                case SpatialUnderstanding.ScanStates.Finishing:
```

```
            CurrentState = State.Finalizing;
            break;
        case SpatialUnderstanding.ScanStates.Done:
            SurfaceMaterial = defaultSurfaceMaterial;
            CurrentState = State.Finished;
            break;
    }
}
```

Instead of disabling the processed surface mesh renderer, we update its `Material`. Before continuing with the code, lets assign the appropriate `Material` to the `defaultSurfaceMaterial` and `scanningSurfaceMaterial` variables.

Back in the Editor, select the `SpatialUnderstanding` GameObject from the **Hierarchy** panel and, within the **Inspector** panel of the **Play Space Scanner** pane, click on the **Default Surface Material** object selector (right-most circle button), search, and select the **Occlusion** material. Once selected, click on the **Scanning Surface Material** object selector, search, and select the `SpatialUnderstandingSurface` material.

With that now done, return to Visual Studio and let's continue fleshing out our `PlaySpaceScanner` class. The next method we will implement is the `Scan` method. Find the method you declared before and make the following amendments:

```
public void Scan()
{
    CurrentState = State.Scanning;

    if (!SpatialMappingManager.Instance.IsObserverRunning())
    {
        SpatialMappingManager.Instance.StartObserver();
    }

    if (SpatialUnderstanding.Instance.AllowSpatialUnderstanding &&
    SpatialUnderstanding.Instance.ScanState ==
    SpatialUnderstanding.ScanStates.None)
    {
        SpatialUnderstanding.Instance.RequestBeginScanning();
    }
}
```

When the `SceneController` calls the `Start` method, we start the `SurfaceObserver` and `SpatialUnderstanding` processes and then constantly poll for updates from the `SpatialUnderstanding` instance to check whether we have sufficiently scanned the environment. We do this within the `Update` method. Let's add this now:

```
void Update()
{
    if(CurrentState == State.Scanning)
    {
        IntPtr statsPtr =
SpatialUnderstanding.Instance.UnderstandingDLL.GetStaticPlayspaceStatsPtr()
;
        if
(SpatialUnderstandingDll.Imports.QueryPlayspaceStats(statsPtr) > 0)
        {
            SpatialUnderstandingDll.Imports.PlayspaceStats stats =
SpatialUnderstanding.Instance.UnderstandingDLL.GetStaticPlayspaceStats();

            if ((stats.TotalSurfaceArea > minAreaForComplete) ||
                (stats.HorizSurfaceArea >
                minHorizontalAreaForComplete) ||
                (stats.WallSurfaceArea >
                minVerticalAreaForComplete))
            {
            CurrentState = State.ReadyToFinish;
            }
        }
    }
}
```

If currently scanning, we query for the current playspace statistics from the `SpatialUnderstanding` library and compare these with the specified criteria to determine whether we have scanned enough. The first call, `SpatialUnderstanding.Instance.UnderstandingDLL.GetStaticPlayspaceStatsPtr()`, is used to create a pointer to the current statistics, if not already created, and binds the values to `reusedPlayspaceStats`. The `reusedPlayspaceStats` class is returned with the next call containing metadata about our environment, such as the horizontal and vertical surface area, number of floors, platforms, and more.

The core of the `SpatialUnderstanding` library was written in C++ (unmanaged); memory-managed, data types, and so on, are handled differently between managed and unmanaged code. One way of communicating between the two is using pointers (IntPtr). Refer to Microsoft's page `An Overview of Managed/Unmanaged Code Interoperability` to learn more about unmanaged and managed Interoperability.

If our criteria are satisfied, then we will update the state to `ReadyToFinish`. In this state, we push the responsibility to the user to signal when to finish scanning. For this signal, we are expecting the air tap gesture but, currently, we haven't registered for these events. Let's fix this now. Add the following code to the `Instance_ScanStateChanged` method:

```
void Instance_ScanStateChanged()
{
    switch (SpatialUnderstanding.Instance.ScanState)
    {
        case SpatialUnderstanding.ScanStates.Scanning:
            InteractionManager.SourcePressed += OnAirTap;
            SurfaceMaterial = scanningSurfaceMaterial;
            CurrentState = State.Scanning;
            break;
        case SpatialUnderstanding.ScanStates.Finishing:
            InteractionManager.SourcePressed -= OnAirTap;
            CurrentState = State.Finalizing;
            break;
        case SpatialUnderstanding.ScanStates.Done:
            SurfaceMaterial = defaultSurfaceMaterial;
            CurrentState = State.Finished;
            break;
    }
}
```

When scanning, we register for the air tap gesture and unregister when we exit. When we detect the air tap, we will request the `SpatialUnderstanding` library to finish scanning, which will update its `ScanState` to `Finishing` and then, finally, `Done`. Add the gesture delegate to `PlaySpaceScanner`, as shown:

```
void OnAirTap(InteractionSourceState state)
{
    if(CurrentState != State.ReadyToFinish ||
    SpatialUnderstanding.Instance.ScanStatsReportStillWorking)
    {
        return;
    }
```

```
        SpatialUnderstanding.Instance.RequestFinishScan();
    }
```

That concludes our `PlaceSpaceScanner` class. When the scene transitions into the scanning state, it will begin scanning the environment. Once sufficiently scanned, the user will be able to air tap to finalize the scan. Once finished, the scene will transition into the placing state, which is the topic of the next section.

The following image shows the result of a scan of a hotel room performed by the `SpatialUnderstanding` library:

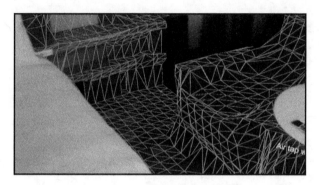

To find locations to place objects, we will be using the object placement solver of the `SpatialUnderstanding`, passing in a query made up of a set of rules and constraints. The object placement solver then searches the playspace to find the most suitable place. It's worth noting that places are persisted until explicitly removed. This allows for multi-object placement avoid having holograms overlap each other. The query consists of the following parts:

- `PlacementType` defines the type of surface to place the hologram on. Some examples include `Place_OnFloor`, `Place_OnWall`, `Place_UnderFurnitureEdge`, and many others.
- One or more `ObjectPlacementRule` define hard rules, meaning that the place cannot violate them. Some examples of placement rules include `Rule_AwayFromPosition`, `Rule_AwayFromWalls`, and `Rule_AwayFromOtherObject`.
- One or more `ObjectPlacementConstraint` are like the object placement rules but are considered soft, meaning that the place is not required to satisfy the constraint but likely to increase its chance of being selected. Some examples include `Constraint_NearPoint`, `Constraint_NearCenter`, `Constraint_AwayFromWall`, and may others.

In the previous section, we decoupled the criteria from the class. In this section, we will take a simpler approach, helping keep the focus on working with the `SpatialUnderstanding` library rather than the supplementary code. Jump back into Visual Studio and open the `PlaySpacePlaceFinder` script; we will start by adding the class variables and properties:

```
public enum States
{
    None,
    Processing,
    Finished
}

public GameObject prefab;

public float distanceFromUser = 0.6f;

public bool Finished { get; private set; }

public States State { get; private set; }

bool solverInitialized = false;

List<SpatialUnderstandingDllObjectPlacement.ObjectPlacementResult>
queryPlacementResults = new
List<SpatialUnderstandingDllObjectPlacement.ObjectPlacementResult>
();
```

Searching can be computationally expensive and, therefore, if offloaded to a separate thread. To help manage the process, we store the current state in the `State` property.

Before being used, the object placement solver needs to be initialized. We use the variable `solverInitialized` to flag whether this step has been performed or not, and initialize before any queries are performed. `prefab` is the prefab of the hologram we want to place. The variable `distanceFromUser` will be a value assigned to a rule added to our query. Our last variable is a list used to store the results of the query. `ObjectPlacementResult` stores the name of the query, position, orientation, and bounds that we will use to place and orientate our hologram.

Initialization is simply a matter on calling
`SpatialUnderstandingDllObjectPlacement.Solver_Init()`. We wrap this in a
method to handle logic of only calling it once and handling cases where the initialization
fails; add the following method to the `PlaySpacePlaceFinder` class:

```
bool InitializeSolver()
{
    if (solverInitialized ||
    !SpatialUnderstanding.Instance.AllowSpatialUnderstanding)
    {
        return solverInitialized;
    }

    if (SpatialUnderstandingDllObjectPlacement.Solver_Init() > 0)
    {
        solverInitialized = true;
    }

    return solverInitialized;
}
```

As mentioned earlier, the places found (the results of the previous queries) are persisted,
allowing for subsequent queries to be made that consider the currently occupied places.
Because we are only placing a single hologram, we will clear any reversed places and the
previous results each time we make a query. Add the following method to handle this:

```
void Reset()
{
    queryPlacementResults.Clear();

    if (SpatialUnderstanding.Instance.AllowSpatialUnderstanding)
    {
    SpatialUnderstandingDllObjectPlacement.Solver_RemoveAllObjects();
    }
}
```

Here, we are simply removing all the results and clearing all persisted places from the
object placement solver. Let's now revisit the `FindPlaces` method and fill in the details;
add the following code to the method:

```
public void FindPlaces()
{
if (!InitializeSolver())
{
    return;
}
```

```
Reset();

Bounds bounds = GetBoundsForObject(prefab);
Vector3 halfDims = new Vector3(bounds.size.x * 0.5f,
bounds.size.y * 0.5f, bounds.size.z * 0.5f);

        SpatialUnderstandingDllObjectPlacement.ObjectPlacementDefinition
placementDefinition =
SpatialUnderstandingDllObjectPlacement.ObjectPlacementDefinition.Create_OnF
loor(halfDims);
        List<SpatialUnderstandingDllObjectPlacement.ObjectPlacementRule>
placementRules = new
List<SpatialUnderstandingDllObjectPlacement.ObjectPlacementRule>() {
SpatialUnderstandingDllObjectPlacement.ObjectPlacementRule.Create_AwayFromP
osition(Camera.main.transform.position, distanceFromUser),
                                    };
        AsyncRunQuery(placementDefinition, placementRules);
    }
```

In this method, once initialized and reset, we obtain the bounds of the prefab and, because the object placement solver is expecting half dimensions, divide each axis in half. We next construct our query parameters by creating a placement definition and list of placement rules. To make life easier, the `ObjectPlacementDefinition` and `ObjectPlacementDefinition` structures expose a set of static builder methods to simplify the process of building a query, which we make use of. We then pass these through to the `AsyncRunQuery` method. The process can be computationally expensive, so the task should be ideally offloaded from the main thread, which is the intention of using this method. Let's add this now:

```
bool
AsyncRunQuery(SpatialUnderstandingDllObjectPlacement.ObjectPlacementDefinit
ion placementDefinition,
        List<SpatialUnderstandingDllObjectPlacement.ObjectPlacementRule>
placementRules = null,
List<SpatialUnderstandingDllObjectPlacement.ObjectPlacementConstraint>
placementConstraints = null)
    {
#if UNITY_WSA && !UNITY_EDITOR
        System.Threading.Tasks.Task.Run(() =>
            {
                RunQuery(placementDefinition, placementRules,
placementConstraints);
            }
        );

        return true;
#else
```

```
            return RunQuery(placementDefinition, placementRules,
placementConstraints);
#endif
    }
```

The most significant part of the code is not the code itself by the preprocessor directives. Here, we run the query on a separate thread if running on the HoloLens device, otherwise we fall back to a synchronized flow, both passing the query down to the RunQuery method that is responsible for handling the query. Let's now implement this method:

```
bool
RunQuery(SpatialUnderstandingDllObjectPlacement.ObjectPlacementDefinition
placementDefinition,
        List<SpatialUnderstandingDllObjectPlacement.ObjectPlacementRule>
placementRules = null,
List<SpatialUnderstandingDllObjectPlacement.ObjectPlacementConstraint>
placementConstraints = null)
{
if (SpatialUnderstandingDllObjectPlacement.Solver_PlaceObject( this.name,
SpatialUnderstanding.Instance.UnderstandingDLL.PinObject(placementDefinitio
n),
            (placementRules != null) ? placementRules.Count : 0,
            ((placementRules != null) && (placementRules.Count > 0)) ?
SpatialUnderstanding.Instance.UnderstandingDLL.PinObject(placementRules.ToA
rray()) : IntPtr.Zero,
            (placementConstraints != null) ? placementConstraints.Count
: 0,
            ((placementConstraints != null) &&
(placementConstraints.Count > 0)) ?
SpatialUnderstanding.Instance.UnderstandingDLL.PinObject(placementConstrain
ts.ToArray()) : IntPtr.Zero,
SpatialUnderstanding.Instance.UnderstandingDLL.GetStaticObjectPlacementResu
ltPtr()) > 0)
    {
            SpatialUnderstandingDllObjectPlacement.ObjectPlacementResult
placementResult =
SpatialUnderstanding.Instance.UnderstandingDLL.GetStaticObjectPlacementResu
lt();

queryPlacementResults.Add(placementResult.Clone() as
SpatialUnderstandingDllObjectPlacement.ObjectPlacementResult);

return true;
}

State = States.Finished;

return true;
```

```
    }
```

Here, we are passing the query to the object placement **Dynamic Link-Library** (DLL) wrapper `SpatialUnderstandingDllObjectPlacement` to perform the query. As mentioned earlier, because the underlying library is written in C++ and complied as an unmanaged DLL, the parameters are needmarshaling. If the query was successful, we will obtain the results and update the state to `Finished`.

Our final method for querying is `GetBoundsForObject` used by the `FindPlaces` method to get the bounds of `prefab`. This method returns the attached collider bounds if one exists, otherwise creating bounds encapsulating all the `renderers`:

```
Bounds GetBoundsForObject(GameObject prefab)
{
    Bounds? bounds = null;

    if (prefab.GetComponent<Collider>() != null)
    {
        bounds = prefab.GetComponent<Collider>().bounds;
    }
    else
    {
        Renderer[] renderers =
        prefab.GetComponentsInChildren<Renderer>();
        foreach (Renderer renderer in renderers)
        {
            if (bounds.HasValue)
            {
                bounds.Value.Encapsulate(renderer.bounds);
            }
            else
            {
                bounds = renderer.bounds;
            }
        }
    }

    return new Bounds(bounds.HasValue ? Vector3.zero :
    bounds.Value.center, bounds.HasValue ? bounds.Value.size :
    Vector3.zero);
}
```

Opportunity to refactor: because we have seen this code twice, we should ask ourselves where we can move it to make it accessible by other classes. One option is some utility class; the other is an extension method on the `GameObject` class, which would be my preferred choice.

Once the state has been updated to `Finished`, `SceneController` calls the `PlaceGameObjects` method on this class, so let's implement this now. Find the `PlaceGameObjects` method and add the following code:

```
public bool PlaceGameObjects()
{
    if(queryPlacementResults.Count == 0)
    {
        return false;
    }
    var objectPlacementResult = queryPlacementResults.First();
    GameObject go = Instantiate(prefab);
    go.transform.position = objectPlacementResult.Position;
    go.transform.up = objectPlacementResult.Up;

    return true;
}
```

At this point, all the hard work has been done. Our final task was to retrieve and use the results from the query to position and orientate our hologram. With this method now complete, we have finished placement using the `SpatialUnderstanding` library.

User placement versus procedural placement

In this chapter, we learned how to better understand the environment and use this information to place our bin with little user intervention. By contrast, in the previous chapter, we had the user explicitly place the bin. So, which one should you use? I would argue, both. The real value is in being able to assist the user in achieving their task, in this instance, placing the bin. The interaction paradigm and technology for MR is still in its infancy and, at times, can be cumbersome to use. By assisting the user in performing tasks, we can make the user more efficient and improve the user experience, similar to how auto-compete improves the writing experience on your smartphone. An concrete example of how we can achieve this with placement is by offering the user a few suitable places and giving them, the user, control to make the final decision.

With our bin now placed, it's time to finish the game by allowing the user toss the paper. At this point, you can use either of the previous two approaches discussed to place the bin.

Nothing but net - throwing paper

As we did in the previous chapter, we will implement throwing by tracking the user's hands and using the tracked direction and displacement to influence the papers flight path, but unlike the previous chapter, we won't be implementing this from scratch, but we'll take advantage of a script that comes bundled with HoloTookit, namely, the `GestureManager` script.

`GestureManager` encapsulates a lot of the logic for detecting *tap* and *manipulation* gestures and managing their state. Here, we are mainly interested in the manipulation gesture. The manipulation gesture is activated when the user holds their finger down and, when activated, the user's hand is tracked with its position being updated and made accessible.

Let's get started by jumping back into the Unity Editor and add this script to the scene. `GestureManager` requires that `GazeManager` is attached to the same GameObject. We will attach it to our `Gaze` GameObject. With the scene open, select the `Gaze` GameObject from within the **Hierarchy** panel, click on the **Add Component** button in the **Inspector** panel, enter, and select `GestureManager` to add to the GameObject. To make use of `GestureManager`, we will implement our own class. Let's do this now. With the `Gaze` GameObject still selected, click on the **Add Component** button in the **Inspector** panel, select **New Script**, and enter the name `PlayerInputController`. This script will be responsible for translating the output received from `GestureManager` into user actions related to our game, such as tossing the paper. Double-click on the `PlayerInputController` script from within the **Project** panel to open it in Visual Studio.

Our approach will resemble what we did in the previous chapter, obviously requiring less code because the tracking is now managed by `GestureManager`, freeing us to concentrate on the application's logic. We will start by registering for the `OnManipulationStarted`, `OnManipulationCompleted`, and `OnManipulationCanceled` events. When we detect a manipulation starting, we will instantiate the paper prefab and start tracking until the gesture is completed or canceled. If canceled, we will simply destroy the instantiated prefab, otherwise we will toss the paper forward in line with the user's hand direction and displacement. Let's start by adding the properties and variables, remembering to include the namespace `HoloToolkit.Unity`. Within the `PlayerInputController` class, add the following to the code:

```
public GameObject PaperPrefab;
```

```
public Vector3 FingertipsOffset = new Vector3(0, 0.053f, 0.01f);

GameObject trackedGameObject;

Vector3 trackingStartPosition = Vector3.zero;

public bool CanTrack
{
    get
    {
        return SceneController.Instance.CurrentState ==
SceneController.State.Playing;
    }
}

public bool IsTracking
{
    get
    {
        return CanTrack &&
GestureManager.Instance.ManipulationInProgress;
    }
}
```

Most of this will look familiar to you, assuming you have read the previous chapter. PaperPrefab is the prefab we will instantiate and throw. trackedGameObject holds reference to the currently tracked object. FingertipsOffset is the offset we apply to the position of the trackedGameObject GameObject when being tracked. We are slightly offsetting the paper from the tracked position because tracking centers at the center of the user's palm, rather than their fingertips, where the user would expect the paper to be held. trackingStartPosition is set when OnManipulationStarted is called and is used to determine how far and what direction the user is trying to throw the paper. The CanTrack and IsTracking properties influence the flow of our code, CanTrack returns true if the application is in a valid state for the user to start tossing paper, while the IsTracking property returns true if we are currently tracking and is used in the Update method when updating the position of trackedGameObject. Let's now turn our attention to the methods that will be responsible for creating, updating, and canceling tracking. Add the following methods to the PlayerInputController class:

```
void CreateAndStartTrackingGameObject()
{
    if (trackedGameObject == null)
    {
        trackedGameObject = Instantiate(PaperPrefab);
        trackedGameObject.GetComponent<Rigidbody>().useGravity =
```

```
            false;
            trackedGameObject.GetComponent<Collider>().enabled = false;
            trackedGameObject.transform.position =
            GestureManager.Instance.ManipulationPosition +
            Camera.main.transform.TransformVector(FingertipsOffset);

            trackingStartPosition =
            trackedGameObject.transform.position;
        }
    }

    void UpdateTrackedObject()
    {
        if (trackedGameObject != null)
        {
            trackedGameObject.transform.position =
            GestureManager.Instance.ManipulationPosition +
            Camera.main.transform.TransformVector(FingertipsOffset);
        }
    }

    void ThrowTrackedGameObject()
    {
        trackedGameObject.GetComponent<Rigidbody>().useGravity = true;
        trackedGameObject.GetComponent<Collider>().enabled = true;
        trackedGameObject.GetComponent<Rigidbody>().velocity =
        velocity;

        trackedGameObject = null;
    }

    void DestoryTrackedGameObject()
    {
        GameObject.Destroy(trackedGameObject);
        trackedGameObject = null;
    }
```

We start with `CreateAndStartTrackingGameObject`; this method is called when the event `OnManipulationStarted` is fired and is responsible for instantiating `PaperPrefab`, updating its position and turning off the gravity of the `Collider`, as we don't want it to react with our environment while being dragged.

The next method, `UpdateTrackedGameObject`, is called from the `Update` method and is responsible for updating the instantiated GameObject's position using the `ManipulationPosition` property of the `GestureManager` class. `ThrowTrackedGameObject` is called once the manipulation gesture is completed, signaled by the event `OnManipulationCompleted`. Here, we are just enabling gravity and the collider, so it interacts to the environment. We will return to this method to add a velocity, allowing the user to toss the paper. The last of these methods is `DestoryTrackedGameObject`. As you might suspect, this method is called when the event `OnManipulationCanceled` is fired and simply destroys the currently tracked object.

With these methods in place, let's register the `GestureManager` events to the appropriate delegates. Within the `Start` method, add the following code:

```
void Start ()
{
    GestureManager.Instance.OnManipulationStarted +=
    Instance_OnManipulationStarted;
    GestureManager.Instance.OnManipulationCompleted +=
    Instance_OnManipulationCompleted;
    GestureManager.Instance.OnManipulationCanceled +=
    Instance_OnManipulationCanceled;
}
```

Now add the following delegates, wiring up the events to our methods:

```
void
Instance_OnManipulationStarted(UnityEngine.VR.WSA.Input.InteractionSourceKi
nd sourceKind)
    {
        if(CanTrack)
        {
            CreateAndStartTrackingGameObject();
        }
    }

    void
Instance_OnManipulationCompleted(UnityEngine.VR.WSA.Input.InteractionSource
Kind sourceKind)
    {
        ThrowTrackedGameObject();
    }

    void
Instance_OnManipulationCanceled(UnityEngine.VR.WSA.Input.InteractionSourceK
ind sourceKind)
    {
```

```
        DestoryTrackedGameObject();
    }
```

No surprises here; we are simply calling the appropriate method for each of the events. With this in place, we almost have everything in place to start tracking. The last method we need to wire up is `UpdateTrackedGameObject`. This will be called from within the `Update` method. Add the following to the `Update` method:

```
void Update () {
    if (IsTracking)
    {
        UpdateTrackedGameObject();
    }
}
```

Now is a good time to stop and test out tracking, but before we test, we first need to wire up the paper prefab to `PlayerInputController`. Jump back into the Editor and select the `Gaze` GameObject from the **Hierarchy** panel. Now, from the **Project** panel's search bar, enter in `CrumbledPaper` and drag the prefab onto the **Paper Prefab** field of `PlayerInputController` in the **Inspector** panel. Once finished, launch the **Build Settings** dialog by selecting the menu item **File** | **Build Settings** and click on **Build** button to build the project.

Our final task is to implement tossing. Similar to the previous chapter, we will fake this using trial and error. Back within Visual Studio, make the following amendments to `ThrowTrackedGameObject` of the `PlayerInputController` class:

```
void ThrowTrackedGameObject()
{

    const float minVelocity = 0.09f;
    const float maxVelocity = 0.3f;

    Vector3 displacement = trackedGameObject.transform.position -
    trackingStartPosition;
    Vector3 direction = displacement.normalized;

    Vector3 velocity = Camera.main.transform.forward * 0.2f;

    if(displacement.magnitude > minVelocity)
    {
        velocity += Camera.main.transform.forward * 2.5f;

        velocity += Mathf.Min(maxVelocity, displacement.magnitude /
        maxVelocity) * direction * 4.0f;
    }
```

```
trackedGameObject.GetComponent<Rigidbody>().useGravity = true;
trackedGameObject.GetComponent<Collider>().enabled = true;
trackedGameObject.GetComponent<Rigidbody>().velocity =
velocity;

trackedGameObject = null;
}
```

As we did in the previous chapter, we are influencing the user's throw based on how far they have moved their hand after initiating the gesture. In this example, we only apply significant velocity if the user has moved their hands beyond a preset threshold. If this distance is satisfied, we add a default velocity based on the user's facing direction along with some additional force based on the direction and distance the user's hand has moved while being tracked. With this method now complete, this concludes this chapter.

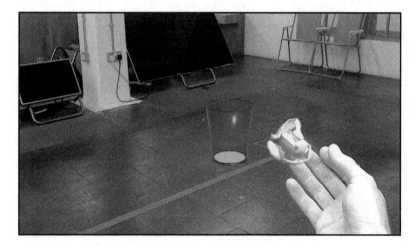

Summary

Once again, congratulations for getting through the chapter. Despite implementing the same project, we have covered a lot. Throughout this chapter, we looked at how we can leverage the open source HolotToolkit project to accelerate our development. We also looked in depth at how to better understand the scanned environment, allowing us to make use of it. We emphasized the importance of user feedback and discussed briefly the important concept of intelligently assisting the user to improve the user experience. We will continue this theme in the next chapter, where we look at voice as our main input.

6
Interacting with Holograms Using Unity

HoloLens, and **mixed reality** (**MR**) applications in general, offer us a glimpse into how we will soon be interacting with our computers. It will be an era where we are freed from the desk, an era where computing is brought into our world--a continuation of what the Smartphone has started. However, with this shift, we will lose some of the comforts and norms we have become accustomed to, such as using dedicated input devices that we use to interact with our computers. This loss means that we need alternative forms of input and output, either by starting from scratch or by adopting the existing forms. The latter, using an existing language, is commonly referred to as **Natural User Interfaces** (**NUI**) and is the path that HoloLens has taken by making its primary forms of input gaze, gesture, and voice; this is the primary focus of this chapter.

This chapter's examples are based on a fictional use case of having to build an application that projects a holographic robot arm into the user's environment and then allowing them to train it using gestures and voice. We will be starting from a base project where scanning and placement have already been implemented, leaving us to focus on implementing how the user interacts with the hologram, specifically implementing the following:

- Navigation gesture
- Manipulation gesture
- Direct manipulation

Then, we will explore two different approaches of how to integrate voice into our application; the two approaches are as follows:

- **Keywords**: Small phrases that encapsulate the intent using a word or a few words, for example, `Take photo`
- **Grammar rule engine**: Phrases defined using **Speech Recognition Grammar Specification** (**SRGS**), allowing for a more natural dialog

> **Collaborative Robots (also called co-bots)**: Have you ever been tasked with a daunting repetitive task, such as peeling buckets of potatoes for a family gathering. This, although an idiotic example, is driving the industry known as collaborative robots (also called co-bots). These are robots designed to work alongside human workers, assisting them with a variety of tasks. Baxter and Sawyer, from Rethink Robotics, are examples of such co-bots.
>
> Unlike traditional industrial robots that require specialized knowledge and experience to program, Baxter and Sawyer can be trained by demonstration, allowing those with little or no technical background to train and (re)deploy these robots for different tasks. This is the idea that inspired our next example.

We will start our journey with a quick explanation of the example we will be working with, including setting up our environment, before jumping to implement the preceding topic.

Project setup

We will start by getting acquainted with the project; if you have not done so already, clone or download the repository for this book from `https://github.com/PacktPublishing/Microsoft-HoloLens-By-Example`. Once downloaded, launch Unity and load the **Starter** project for this chapter found in the `Chapter6/Starter` directory.

Once loaded, your project should look similar to this:

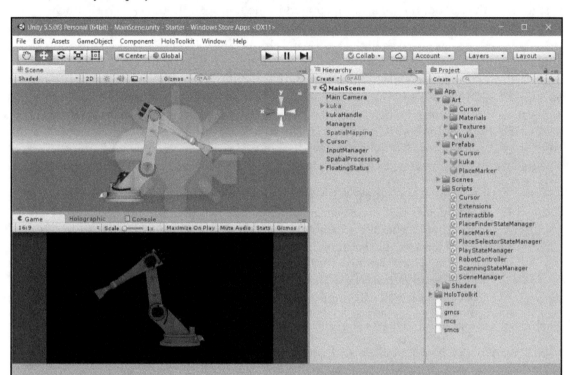

Credit and thanks to *Menagy* for the use of the model used in this project; it is available at http://www.blendswap.com/blends/view/19629.

As mentioned already, scanning and placement have been implemented using the techniques we covered in the last two chapters, leaving us to focus solely on the ways of interacting with the hologram. In this section, we will overview the main components to give you a sense of how the application is structured.

The scene states are managed by the SceneManager script, which hands over control to the PlayStateManager script once the hologram has been placed, and the entry point from where we will begin our journey.

The hologram we will be interacting with is a Kuka robot arm (shown in the preceding image) with the bulk of the logic encapsulated in the `RobotController` script attached to the root GameObject of the robot. The methods we are interested in are `Rotate` and `MoveIKHandle`. `Rotate` takes in the name of a child GameObject (obtained from a collision), Euler angle containing the change in rotation, and origin of rotation (local or world); this method then simply finds the child transform using the name, and applies the specified rotation relative to the specific origin. The `MoveIKHandle` method expects a translation vector, which is then applied to an external target GameObject that the arm will seek using a simple **Inverse Kinematics** (**IK**) solver. When rotating using the `Rotate` method, the IK solver must be disabled and only enabled when manipulating this handle; this can be done by toggling the `solverActive` variable of `RobotController`.

Inverse Kinematics (**IK**) refers to an algorithm used to determine the joint parameters that place the end-effector at a specific location. The algorithm used in this chapter is called **Cyclic Coordinate Descent** (**CCD**) and provides a simple solution by only solving for a single joint at a time, working from the end-effector (the tip of the robot) to the root (base), with each iteration rotating in the direction to reduce the angle distance between the end-effector and target.

Each part is constrained to the axis which it can rotate around.

- Base can be rotated around the world's y axis
- **Arm 1** and **Arm 2** can each be rotated around the world's x axis
- The tool acts as the end-effector of the inverse kinematics chain and rotates the **Base**, **Arm 1**, and **Arm 2** toward its target, the handle, as described earlier

The following figure illustrates these parts, along with the axes they can be manipulated around, starting at the bottom:

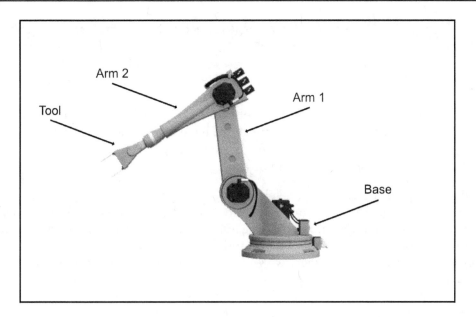

It will be our responsibility, via the PlayStateManager script, to interpret the user's intention using gestures and/or voice, and proxy these to the RobotController. However, before we move on to adding gestures, let's quickly explore the PlayStateManager script to get familiar with it and its nuances. In the Unity editor, look for the script by clicking on the search field of the **Project** panel and entering PlayStateManager; when visible, double-click to open it in Visual Studio.

Currently, the PlayStateManager is concerned with the following:

- Knowing when it's active, is done by registering for the OnStateChangedEvent event in the Start method, which the SceneManager script broadcasts when the state changes. This, in turn, calls the OnStateActiveChanged method. Later on, we can use this method to register for gestures and voice keywords.
- Next, is knowing what the user is currently gazing at so that we know what to modify when the user performs a gesture or voice command. This is done by polling the GazeManager in the LateUpdate method and assigning the current FocusedObject to the equivalent local variable. Within this property, we call OnFocusedObjectChanged, passing in the previous and currently focused GameObject. This method will query the currently focused GameObject for the Interactible component and, if attached, will assign it to the CurrentInteractible property, which simply passes GazeExited on any previously set Interactible and GazeEntered for the CurrentInteractible.

- Finally, `PlayStateManager` manages selecting the `CurrentInteractible`, which essentially locks the `CurrentInteractible` until deselected. Selecting is done by calling `SelectCurrentInteractible` when a `CurrentInteractible` is available, and calling `DeselectCurrentInteractible` to unselect.

 Like the `Update` method, `LateUpdate` is called once every frame, but is called after all the `Update` functions have been called. It is useful for anything that is dependent on something being updated, such as having a camera follow an object.

You will now hopefully have a high-level understanding of the relevant parts for this project, at least enough for us to move on and look at gestures and voice. To help make the concepts more relevant, build and deploy the application in the emulator or device to get a feel of how the application currently works (and that it does work); then, return here to continue with the next section, where we will look at adding our first form of interaction-- gestures.

Interacting with gestures

In the last two chapters, we explored the air-tap and manipulation gestures, including tracking the user's hands using `InteractionManager`; in this section, we will introduce the navigation gesture and revisit the manipulation to give the user full control over the robot hand via the inverse kinematics target, and hand tracking to allow the user to "push" the robot arm around.

Similar to the manipulation gesture, the navigation gesture becomes active when a hold state is detected (when the user performs an air-tap, but holds their finger down), but differs in that it can be locked to a specific axis/axes. It has its result normalized between -1.0 to 1.0 based on the offset the user's hand is from the position that initiated the gesture, akin to a joystick.

Let's start by adding the navigation gesture. In the Unity editor, open the `PlayStateManager` script in Visual Studio, if not already open, by entering `PlayStateManager` into the search field of the **Project** panel, and double-clicking to open it in Visual Studio.

To get access to the gestures, we need to include the appropriate namespace. At the top of the script, add the `UnityEngine.VR.WSA.Input` namespace; with that now included, let's discuss what we are trying to achieve, and our approach.

The navigation gesture is suited for controlling joints that are constrained to a single axis, while the manipulation gesture is better suited to manipulate the handle that the robot arm will follow when `solverActive` is set to true. These gestures share a lot in common, from the activation to the actual gesture and therefore we cannot use these two gestures at the same time. For this reason, we will have a single gesture recognizer active at any given time and switch between them based on the user context, or in our case, gaze.

As mentioned earlier, the `PlayStateManager` keeps track of the currently focused GameObject (based on the users gaze), and when set, it will check whether a `Interactible` component is attached, and if so, this component will assign it to the `CurrentInteractible` property, which in turn calls the `OnCurrentInteractibleChanged` method. `Interactible` encapsulates metadata about a specific joint, such as the type of interaction, axis it can rotate on, and some visual aesthetics information for when it's in focus or is selected. We will hook into the `OnCurrentInteractibleChanged` method to test how many axes the joint is free to move, and start the appropriate gesture recognizer. Let's put this into practice by first declaring the variables and properties we need; within the body of the `PlayStateManager` class, add the following lines of code:

```
private GestureRecognizer navigationRecognizer;

private GestureRecognizer manipulationRecognizer;

private GestureRecognizer _activeRecognizer;

public GestureRecognizer ActiveRecognizer
{
get { return _activeRecognizer; }
private set
{
if(_activeRecognizer == value) { return; }

if(_activeRecognizer != null)
{
_activeRecognizer.StopCapturingGestures();
}

_activeRecognizer = value;

if(_activeRecognizer != null)
{
_activeRecognizer.StartCapturingGestures();
}
}
}
```

We will instantiate two recognizers, one specifically for handling manipulation gestures and the other for handling navigation gestures. The currently active recognizer will be assigned to the `ActiveRecognizer` property, which is responsible for stopping the currently active recognizer (if any) and starting the new one. As we discovered in the last chapter, `GestureRecognize` uses the delegation pattern to communicate updates; let's wire these up for both recognizers. Within the `Start` method, add the following code:

```
navigationRecognizer = new GestureRecognizer();
navigationRecognizer.SetRecognizableGestures(GestureSettings.NavigationX |
GestureSettings.NavigationY);
navigationRecognizer.NavigationStartedEvent +=
NavigationRecognizer_NavigationStartedEvent;
navigationRecognizer.NavigationUpdatedEvent +=
NavigationRecognizer_NavigationUpdatedEvent;
navigationRecognizer.NavigationCompletedEvent +=
NavigationRecognizer_NavigationCompletedEvent;
navigationRecognizer.NavigationCanceledEvent +=
NavigationRecognizer_NavigationCanceledEvent;

manipulationRecognizer = new GestureRecognizer();
manipulationRecognizer.SetRecognizableGestures(GestureSettings.Manipulation
Translate);
manipulationRecognizer.ManipulationStartedEvent +=
ManipulationRecognizer_ManipulationStartedEvent;
manipulationRecognizer.ManipulationUpdatedEvent +=
ManipulationRecognizer_ManipulationUpdatedEvent;
manipulationRecognizer.ManipulationCompletedEvent +=
ManipulationRecognizer_ManipulationCompletedEvent;
manipulationRecognizer.ManipulationCanceledEvent +=
ManipulationRecognizer_ManipulationCanceledEvent;
```

Here, we are instantiating each of our recognizers, setting the gestures we are interested in, and then simply assigning our yet-to-be defined delegates. Let's now stub out those delegate methods; add the following code within the `PlayStateManager` class:

```
void ManipulationRecognizer_ManipulationStartedEvent(InteractionSourceKind
source, Vector3 cumulativeDelta, Ray headRay)
 {
 }

void ManipulationRecognizer_ManipulationUpdatedEvent(InteractionSourceKind
source, Vector3 cumulativeDelta, Ray headRay)
 {
 }

void
ManipulationRecognizer_ManipulationCompletedEvent(InteractionSourceKind
```

```
source, Vector3 cumulativeDelta, Ray headRay)
 {
 }

void ManipulationRecognizer_ManipulationCanceledEvent(InteractionSourceKind
source, Vector3 cumulativeDelta, Ray headRay)
 {
 }

void NavigationRecognizer_NavigationStartedEvent(InteractionSourceKind
source, Vector3 normalizedOffset, Ray headRay)
 {
 }

void NavigationRecognizer_NavigationUpdatedEvent(InteractionSourceKind
source, Vector3 normalizedOffset, Ray headRay)
 {
 }

void NavigationRecognizer_NavigationCompletedEvent(InteractionSourceKind
source, Vector3 normalizedOffset, Ray headRay)
 {
 }

void NavigationRecognizer_NavigationCanceledEvent(InteractionSourceKind
source, Vector3 normalizedOffset, Ray headRay)
 {
 }
```

We will flesh these out very soon, but before doing so, let's add the code that manages the currently active gesture recognizer. Find the PlayStateManager_OnCurrentInteractibleChanged method and add the following code:

```
if(CurrentInteractible == null)
 {
 ActiveRecognizer = null;
 return;
 }

 if(CurrentInteractible.interactionType ==
Interactible.InteractionTypes.Manipulation)
 {
 ActiveRecognizer = manipulationRecognizer;
 }
 else
 {
 ActiveRecognizer = navigationRecognizer;
```

```
    }
```

As mentioned in the preceding code snippet, this method is called when the
CurrentInteractible property is updated, which is an effect of the currently focused
GameObject being updated. When this is updated, we first see whether we have reference
to an Interactible; if not, we set the ActiveRecognizer property to null and return;
otherwise, we set it to the appropriate recognizer based on the interaction type of
the Interactible.

With the appropriate GestureRecognizer started, we can now return to the delegates to
flesh them out, starting with the navigation gesture.

When the gesture is started, we want to lock the currently focused joint; we do this
by calling SelectCurrentInteractible; similarly, when the gesture is complete or
cancelled, we will unlock via the DeselectCurrentInteractible method. Make the
following changes to the navigation's started, complete, and cancelled delegates:

```
void NavigationRecognizer_NavigationStartedEvent(InteractionSourceKind
source, Vector3 normalizedOffset, Ray headRay)
  {
    SelectCurrentInteractible();
  }

void NavigationRecognizer_NavigationCompletedEvent(InteractionSourceKind
source, Vector3 normalizedOffset, Ray headRay)
  {
    DeselectCurrentInteractible();
  }

void NavigationRecognizer_NavigationCanceledEvent(InteractionSourceKind
source, Vector3 normalizedOffset, Ray headRay)
  {
    DeselectCurrentInteractible();
  }
```

Finally, we want the gesture updates to affect the currently selected joint. To determine
what axis to use from the result, we must first determine the relative axis of the robot,
keeping in mind that the navigation gesture is only for joints that are constrained to a single
axis. This axis will determine what axis, to use of the gesture's result; add the following
code to the navigation's updated delegate:

```
void NavigationRecognizer_NavigationUpdatedEvent(InteractionSourceKind
source, Vector3 normalizedOffset, Ray headRay)
  {
    if(Mathf.Abs(CurrentInteractible.interactionAxis.x) > 0)
    {
```

```
Robot.Rotate(CurrentInteractible.name, CurrentInteractible.interactionAxis
* normalizedOffset.y);
}
else
{
Robot.Rotate(CurrentInteractible.name, CurrentInteractible.interactionAxis
* -normalizedOffset.x);
}
}
```

This finishes the code required to allow the user to interact with the joints constrained to a single axis; let's now turn our attention to the manipulation gesture and the associated delegates.

We will follow a similar pattern as followed earlier, but with some additional blocks of code to handle calculating the change in distance and turning the inverse kinematics solver on and off. Start by adding the following variable to the PlayStateManager class:

```
private Vector3 previousCumulativeDelta = Vector3.zero;
```

We store the previous value returned by the manipulation gesture recognizer to calculate the change for each update. The parameter given from the gesture recognizer is, as the name suggests, a cumulative change in position from when the gesture started, rather than a displacement from the last frame, which is more useful for what we are trying to achieve. Make the following changes to the started, completed, and cancelled manipulation delegate methods:

```
void ManipulationRecognizer_ManipulationStartedEvent(InteractionSourceKind
source, Vector3 cumulativeDelta, Ray headRay)
{
SelectCurrentInteractible();
Robot.solverActive = true;
previousCumulativeDelta = cumulativeDelta;
}

void
ManipulationRecognizer_ManipulationCompletedEvent(InteractionSourceKind
source, Vector3 cumulativeDelta, Ray headRay)
{
DeselectCurrentInteractible();
Robot.solverActive = false;
}

void
ManipulationRecognizer_ManipulationCanceledEvent(InteractionSourceKind
source, Vector3 cumulativeDelta, Ray headRay)
{
```

```
DeselectCurrentInteractible();
Robot.solverActive = false;
}
```

As we did earlier, we perform the following steps when the gesture starts:

1. Lock the currently set `Interactible` by calling `SetCurrentInteractible` and unlock when the gesture finishes by calling `DeselectCurrentInteractible`.
2. Store the `cumulativeDelta` parameter to be used for calculating the offset when the gesture notifies us of an update.
3. Finally, enable and disable the robot arm's IK solver, for the reasons discussed.

The final delegate is to handle the actual update, which simply calculates the displacement and passes this to the `MoveIKHandle` method of the `RobotController` to reposition the handle; add the following code to the manipulations updated delegate:

```
void ManipulationRecognizer_ManipulationUpdatedEvent(InteractionSourceKind
source, Vector3 cumulativeDelta, Ray headRay)
{
Vector3 delta = new Vector3(cumulativeDelta.x - previousCumulativeDelta.x,
(cumulativeDelta.y - previousCumulativeDelta.y), cumulativeDelta.z -
previousCumulativeDelta.z);
Robot.MoveIKHandle(delta);
previousCumulativeDelta = cumulativeDelta;
}
```

With this now complete, you can gaze and manipulate joints of the robot with gestures. Now is a good time to build and deploy the application in the emulator or device to see it in action:

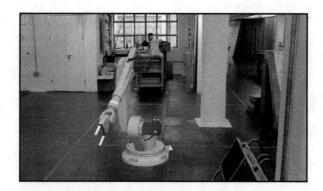

So far in this section, we have seen how we can use the navigation and manipulation gestures to manipulate our robot, either rotating it around a single axis using the navigation gesture. or dragging it around using the manipulation gesture. Before moving on to voice, we will look at how we can go beyond gestures and explore how we might directly approach manipulating the robot, allowing users to virtually "touch" the robot.

In Chapter 4, *Building Paper Trash Ball in Unity*, we saw how to use InteractionManager to track the hands that were visible in the gesture frame, and used this data to allow the user to project virtual paper into the environment. If you played with it, (and I hope you did) you would have found a few shortcomings, namely, the limited tracking range and accuracy. However, despite the technology being relatively limited, it's important to remind ourselves that the HoloLens is a developer device, and this technology will improve vastly within a short time; therefore, in this instance, I would encourage designing for tomorrow, rather than for today.

We will use a lot of what we have already covered in the earlier chapters, but using it in the context of allowing the user to push and prod the robot arm. To avoid bundling too much code into a single class, we will create a separate class that will be responsible for tracking the user's hands along with taking care of the manipulation of the robot arm.

Back in the Unity editor, create a new script by clicking on the **Create** button near the top of the **Project** panel and select **C# Script**; name this script as PlayStateHandTracker, and double-click on the newly created script to open it in Visual Studio. However, before writing any code, let's quickly review our approach.

From Chapter 4, *Building Paper Trash Ball in Unity*, we learned that InteractionManager fires events when a hand (one kind of interaction sources) is discovered, updated, and lost. Using this information, we are able to track the user's hand, but, in order to have it interact with our hologram, we need to have a digital representation of it, such that the robot arm and user's hand are in the same space. For this, we will implement a simple MonoBehaviour class that will be responsible for representing the user's hand in the virtual space and react with the robot when colliding with it. If you inspect the **kuka** GameObject within the editor and drill down its children, you will come across a child with the name **arm_011**; attached to this is the Interactible component and BoxCollider--both of which we will use to detect when the user's hand has come in contact with the robot. When in contact, we will call the MoveIKHandle method of RobotController, as we did for the manipulation gesture, but using the tracked hand's change in position as the displacement. With our approach now fleshed out, let's jump back to Visual Studio and start making amendments to the PlayStateHandTracker class we just created.

Start by adding the `UnityEngine.VR.WSA.Input` namespace to gain access to the class `InteractionManager` and its accompanying types. As mentioned in the preceding section, we need a class to represent the user's hand in virtual space; for this, we will use a simple class that is created when a source is detected and removed when the source is lost. Add the following nested class to our `PlayStateHandTracker` class:

```
public class TrackedHand : MonoBehaviour
{
public uint handId;

public Vector3 translation = Vector3.zero;

void Awake()
{
GameObject go = GameObject.CreatePrimitive(PrimitiveType.Sphere);
go.transform.parent = transform;

transform.localScale = new Vector3(0.1f, 0.1f, 0.1f);
}

public void UpdatePosition(Vector3 position)
{
translation = position - transform.position;
this.transform.position = position;
}
}
```

As mentioned in the preceding code sinppet, the `TrackedHand` will be instantiated when a hand is detected and removed when the hand is lost. We are temporarily creating a Sphere for development and demonstration purposes, which we will remove soon.

Next, we will register (and unregister) for the appropriate events of the `InteractionManager`; add the following code to the `PlayStateHandTracker` class:

```
void Start ()
{
RegisterForInteractionEvents();
}

void OnDestroy()
{
UnregisterFromInteractionEvents();
}

public void RegisterForInteractionEvents()
{
InteractionManager.SourceDetected += InteractionManager_SourceDetected;
```

```
InteractionManager.SourceUpdated += InteractionManager_SourceUpdated;
InteractionManager.SourceLost += InteractionManager_SourceLost;
}

public void UnregisterFromInteractionEvents()
{
InteractionManager.SourceDetected -= InteractionManager_SourceDetected;
InteractionManager.SourceUpdated -= InteractionManager_SourceUpdated;
InteractionManager.SourceLost -= InteractionManager_SourceLost;
}
```

In the preceding code snippet, we are simply registering for the necessary events when the `MonoBehaviour` starts and unregistering when destroyed. Before we implement the delegates, let's write two helper methods that can be used by these delegate methods to handle tracking, and when tracking is lost, add the following code:

```
TrackedHand TrackHand(InteractionSourceState state)
{
if (state.source.kind != InteractionSourceKind.Hand)
{
return null;
}

Vector3 handPosition;
if (!state.properties.location.TryGetPosition(out handPosition))
{
return null;
}

handPosition += state.headRay.direction * 0.1f;

if (!trackedHands.ContainsKey(state.source.id))
{
GameObject go = new GameObject(string.Format("TrackedHand_{0}",
state.source.id));
TrackedHand trackedHand = go.AddComponent<TrackedHand>();
trackedHand.handId = state.source.id;
trackedHand.UpdatePosition(handPosition);

trackedHands.Add(state.source.id, trackedHand);
}

trackedHands[state.source.id].UpdatePosition(handPosition);

return trackedHands[state.source.id];
}

void LostTracking(InteractionSourceState state)
```

```
{
if (trackedHands.ContainsKey(state.source.id))
{
TrackedHand trackedHand = trackedHands[state.source.id];
trackedHands.Remove(state.source.id);
Destroy(trackedHand.gameObject);
}
}
```

The `TrackHand` method will be called when a source is detected and updated; this method is responsible for filtering out any other source that is not a hand. If a hand is found, we try to get the its position using the `TryGetPosition` method of the state's location property, exiting if this fails, otherwise nudging the position toward its `headRay` direction to better align it with the user's hand (set through trial and error). If no tracked hand exists with the source ID, then one is created before calling `UpdatePosition` and passing it the amended `handPosition` variable. The `LostTracking` method simply destroys the associated GameObject, if one exists.

With our helper methods now written, we are just left with calling them via the delegates we have assigned to the `InteractionManager`; add the following code:

```
void InteractionManager_SourceDetected(InteractionSourceState state)
{
TrackHand(state);
}

void InteractionManager_SourceUpdated(InteractionSourceState state)
{
TrackHand(state);
}

void InteractionManager_SourceLost(InteractionSourceState state)
{
LostTracking(state);
}
```

Now, with this implemented, jump back into the Unity editor to attach the script to our `Managers` GameObject. With the editor open, select the `Managers` GameObject from within the **Hierarchy** panel before clicking on the **Add Component** button in the **Inspector** panel and type in `PlayStateHandTracker` to attach our newly created script. You are now ready to build and deploy the application to the device to get a visualization of hand tracking.

We now are able to track the user's hands. In the last part, we will allow the user to "touch" the hologram; the majority of the code is in place, what's remaining is being aware of when the user is touching the robot and handling when the hand comes in contact with it--lucky for us, Unity makes this a breeze. Jump back in the `PlayStateHandTracker` script and within the `TrackedHand` nested class, make the following amendments to the `Awake` method:

```
void Awake()
{
Collider collider = gameObject.AddComponent<SphereCollider>();
collider.isTrigger = true;

Rigidbody rigidBody = gameObject.AddComponent<Rigidbody>();
rigidBody.isKinematic = true;
rigidBody.useGravity = false;

transform.localScale = new Vector3(0.1f, 0.1f, 0.1f);
}
```

Here, we are attaching a `Collider` and `rigidBody`, disabling any effects caused by forces and collisions by setting `isKinematic` to true. With these two components attached, our Gameobject will receive events when our GameObject comes into and out of contact with any other GameObject with a `Collider` attached. The GameObject is notified about these events via the `OnTriggerEnter`, `OnTriggerExit`, and `OnTriggerStay` methods, which is where we will place our code to handle manipulating the robot. Add the following code to the `TrackHand` class:

```
void OnTriggerEnter(Collider other)
{
Interactible interactible = other.transform.GetComponent<Interactible>();
if(interactible != null && interactible.interactionType ==
Interactible.InteractionTypes.Manipulation)
{
PlayStateManager.Instance.Robot.solverActive = true;
trackedInteractible = interactible;
}
}

void OnTriggerExit(Collider other)
{
if(trackedInteractible != null)
{
PlayStateManager.Instance.Robot.solverActive = false;
}
```

```
trackedInteractible = null;
}
```

When a collision occurs, we first determine whether we are colliding with the end-effector of the robot. If so, we assign its `Interactible` component to the `trackedInteractible` local variable and enable the IK solver by setting the `solveActive` of `RobotController` to `true`, and of course, ensuring that we disable it when we exit the collision via the `OnTriggerExit` method. We now have reference to the robot when the user's hand collides with its end-effector. The final thing left to do is to update the handle--the target used by the inverse kinematic chain--as the user moves their hand. Similar to how we handled the manipulation gesture, we will manipulate the handle by passing through the change in position of the user hand. Make the following amendments to the `UpdatePosition` method of the `TrackedHand` class:

```
public void UpdatePosition(Vector3 position)
{
translation = position - transform.position;
this.transform.position = position;

if(trackedInteractible != null)
{
PlayStateManager.Instance.Robot.MoveIKHandle(translation);
}
}
```

With this in place, the user is now able to "touch" the holographic robot arm. Let's see it in action--build and deploy the application to the device, place the robot so that its end-effector is within arm's reach, and reach out to touch it:

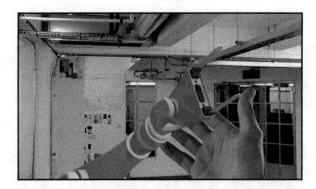

Direct manipulation is what made **Graphical User Interfaces (GUI)** so successful over the **Command Line Interface (CLI)**, but has, until recently, been done through an intermediary device, whether a keyboard and mouse or touchscreen. HoloLens removes this layer, bringing us closer to the content we are concerned about rather than the tooling. In this section, we have explored how we can implement a form of interaction similar to that of a scene from the movie *Minority Report* (2002). In the next section, we will shift our focus from direct manipulation to conversation and explore how we can integrate voice into our application.

Voice

Voice is quickly becoming a prominent way of how we interact with our digital world. During the 2016 Google I/O Keynote, Google CEO Sundar Pichai announced that 20% of all queries on its mobile app and Android devices were made using voice. This shouldn't come as a surprise; advances in speech recognition and understanding coupled with the convenience of voice, makes it an efficient and natural way of interacting with our devices, if used properly.

HoloLens provides a natural fit for voice due to its lack of peripheral devices and contextually situated placement. In this section, we will explore how we can make use of it; we will start by looking at how to use keywords to control the robot (independent phrases that encapsulate a specific task), and then explore how we can allow for more complex forms of voice input, such as dialog and conversational interactions.

Keyword recognizer

Keywords are akin to commands typed in a CLI, in our case, a word or short phrase that encapsulates a specific action; some examples include (all are HoloLens system commands):

- `Select` is used to select the currently focused object
- `Go back` to return to the previous screen
- `Hide` to hide the currently focused object

For each utterance, there is no need for conversing or dialog with the user; the intent is obvious, but what is notable is that they are contextually dependent and multimodal, which means that they are dependent on other forms of input to interpret a user's intent correctly; for example, Select will select the item the user is currently gazing at, where gaze is the other input providing context. Multimodal interfaces provide a more natural form of interaction, something more aligned with how we interact with other people.

HoloLens makes it easy to add voice keywords into your application, and, in this section, we will do just that. We will start by adding the necessary code to listen and handle keywords, and then we will explore how we can create a more natural experience by better understanding the user.

Before committing to any code, let's quickly discuss what we are trying to achieve. However, first a disclaimer--the examples we present here do not adhere to best practices for voice interfaces, but rather, for demonstration purposes only, showing how to integrate voice functional into your application; for a more comprehensive guide to designing voice interfaces, I recommend *Designing Voice User Interfaces* by Cathy Pearl.

As mentioned earlier, keywords are single words or short phrases that denote a specific action. Given that our goal for this example is to allow the user to program (or control) the robot arm, we will simply extend this capability to voice. The following is a list of phrases and their associated actions, we will implement:

Phrases	Action
rotate left, rotate right	When recognized, start rotating the base of our robot in the specified direction
rotate up, rotate down	Start rotating the currently focused arm, if any, in the specified direction; for example, if the user is currently gazing at arm 1, we will infer this to be the arm the user is referring to
rotate arm 1 up, rotate arm 1 down	Start rotating arm 1 in the specified direction
rotate arm 2 up, rotate arm 2 down	Start rotating arm 2 in the specified direction
move up, move down, move forward, move backwards, move left, move right	Start moving the inverse kinematics target (IK handle) in the specified direction
stop	Stop the current action

With our vocabulary now defined, let's turn our attention to putting these concepts into practice. Jump back into the Unity editor, and we will start by defining an abstract class that will act as a contract for the two approaches for integrating voice into our application. This will allow us to easily swap the behavior with minimal changes to our code, known in software engineering as the **strategy pattern**.

Click on the **Create** button within the **Project** panel and select **C# Script**; enter the name `PlayStateVoiceHandler`. We will make this class abstract and implement common functionality that is likely to be shared between the two approaches, two of which will be abstract methods for starting and stopping the handler, some constants for the part names and helper methods to translate directions into something usable. Double-click on the `PlayStateVoiceHandler` script to open it in Visual Studio and make the following amendments:

```
public abstract class PlayStateVoiceHandler : MonoBehaviour {

  sealed public class Direction {
  public const string Left = "Left";
  public const string Right = "Right";
  public const string Up = "Up";
  public const string Down = "Down";
  public const string Forward = "Forward";
  public const string Back = "Back";
  }

  protected const string PART_BASE = "Base";
  protected const string PART_ARM_1 = "Arm 1";
  protected const string PART_ARM_2 = "Arm 2";
  protected const string PART_HANDLE = "Handle";

  public abstract void StartHandler();

  public abstract void StopHandler();

  protected Vector3 GetRotationVector(string direction, float magnitude =
  1f)
  {
  switch (direction)
  {
  case Direction.Up:
  return new Vector3(1f * magnitude, 0, 0);
  case Direction.Down:
  return new Vector3(-1f * magnitude, 0, 0);
  case Direction.Left:
  return new Vector3(0, 0, -1 * magnitude);
  case Direction.Right:
```

```
    return new Vector3(0, 0, 1 * magnitude);
    }

    return Vector3.zero;
    }

    protected Vector3 GetTranslationVector(string direction, float magnitude =
    1f)
    {
    switch (direction)
    {
    case Direction.Up:
    return new Vector3(0, 1f * magnitude, 0);
    case Direction.Down:
    return new Vector3(0, -1f * magnitude, 0);
    case Direction.Left:
    return new Vector3(-1 * magnitude, 0f, 0);
    case Direction.Right:
    return new Vector3(1 * magnitude, 0f, 0);
    case Direction.Forward:
    return new Vector3(0, 0, 1f * magnitude);
    case Direction.Back:
    return new Vector3(0, 0, -1f * magnitude);
    }

    return Vector3.zero;
    }

    }
```

With our base class defined, let's jump back to the Unity editor and implement the class responsible for listening and handling the phrases defined earlier. Click on the **Create** button within the **Project** panel, select **C# Script**, enter the name PSKeywordHandler, and double-click to open it in Visual Studio.

Let's first add the necessary namespaces; add the following lines at the top of the script:

```
using UnityEngine.Windows.Speech;
using System.Linq;
using System.Text.RegularExpressions;
using HoloToolkit.Unity;
```

The UnityEngine.Windows.Speech namespace, as the name suggests, gives us access to the Windows Speech API, which we make extensive use of in the remainder of this chapter. Next, extend the PSKeywordHandler class with PlayStateVoiceHandler, for reasons described already. With housekeeping out of the way, we can now concentrate on implementing the necessary functionalities, starting with listening to a user's utterances and handling the specified phrases.

The class that will perform all the heavy lifting is KeywordRecognizer; we can integrate voice capabilities into our application by simply passing an array of phrases we're interested in (and optionally, a minimum confidence level) and assigning a delegate to handle the results. Let's do this now; add the following code to the StartHandler method:

```
KeywordRecognizer keywordRecognizer;

public override void StartHandler()
{
var keywordCollection = new List<string>();

keywordCollection.Add("rotate left");
keywordCollection.Add("rotate right");

keywordCollection.Add("rotate up");
keywordCollection.Add("rotate down");

keywordCollection.Add("rotate one up");
keywordCollection.Add("rotate one down");

keywordCollection.Add("rotate two up");
keywordCollection.Add("rotate two down");

keywordCollection.Add("move up");
keywordCollection.Add("move down");
keywordCollection.Add("move left");
keywordCollection.Add("move right");
keywordCollection.Add("move forward");
keywordCollection.Add("move back");

keywordCollection.Add("stop");

keywordRecognizer = new KeywordRecognizer(keywordCollection.ToArray(),
ConfidenceLevel.High);
keywordRecognizer.OnPhraseRecognized +=
KeywordRecognizer_OnPhraseRecognized;
keywordRecognizer.Start();
}
```

Being good citizens, we'll also add code to unregister the delegate and stop `KeywordRecognizer`. Once stopped and destroyed, add the following code:

```
public override void StopHandler()
{
if(keywordRecognizer == null)
{
return;
}

keywordRecognizer.OnPhraseRecognized -=
KeywordRecognizer_OnPhraseRecognized;
keywordRecognizer.Stop();
keywordRecognizer.Dispose();
}

void OnDestroy()
{
StopHandler();
}
```

In the `StartHandler` method, we simply create a list of phrases we are interested in capturing, passing them to the constructor of the `KeywordRecognizer`.
When `KeywordRecognizer` recognizes any of these phrases, it will pass the results back to the yet-to-be-defined delegate, along with the `PhraseRecognizedEventArgs` argument. This object includes details such as the following:

- **Text**: The interpreted utterance of the user (in most instances, matching one of your phrases)
- **Confidence**: The level of confidence the `KeywordRecognizer` has in its interpretation of the text
- **phraseStartTime**: The start time when the phrase was first detected
- **phraseDuration**: The time it took for the phrase to be spoken
- **semanticMeanings**: An array of `SemanticMeaning`, semantic properties that have been specified in an SRGS grammar file which is metadata associated to a recognized utterance (something we will cover in the next section)

Finally, we call `Start` to activate `KeyRecognizer`; let's now implement the `OnPhraseRecognized` delegate by adding the following code:

```
void KeywordRecognizer_OnPhraseRecognized(PhraseRecognizedEventArgs args)
{
Debug.LogFormat("Heard {0} ({1})", args.text, args.confidence);
}
```

As mentioned, this delegate is called whenever a specified phrase has been recognized (and is above the given confidence threshold) by `KeywordRecognizer`. Now is a good time to test, build, and deploy to test each of the phrases and to ensure that everything is working.

Our next task is to perform some action when we recognize a phrase. For each phrase, we want to change the state of the robot; this state will be determined by either the phrase in isolation or the phrase within the context of the user's gaze. For this example, we will manage state simply by managing the currently selected direction and part. For example, if the phrase `move left` is recognized, we will set the part to base and direction to left, and continue updating until the state changes.

Let's go ahead and define the types, variables, and properties to manage this state; add the following code to your `PSKeywordHandler` class:

```
public float rotationSpeed = 5.0f;

public float moveSpeed = 5.0f;

public string CurrentDirection { get; set; }

private string _currentPart = null;

public string CurrentPart
{
get
{
return _currentPart;
}
set
{
if(_currentPart != null)
{
if (_currentPart.Equals(PART_HANDLE) &&
(PlayStateManager.Instance.CurrentInteractible != null &&
PlayStateManager.Instance.CurrentInteractible.interactionType !=
Interactible.InteractionTypes.Manipulation))
{
PlayStateManager.Instance.Robot.solverActive = false;
```

```
    }
    }

    _currentPart = value;

    if (_currentPart != null)
    {
    if (_currentPart.Equals(PART_HANDLE))
    {
    PlayStateManager.Instance.Robot.solverActive = true;
    }
    }
    }
    }
```

First, we define a property to track the currently selected part, CurrentPart, and its currently assigned behavior, CurrentDirection, set by parsing the captured keywords. The CurrentPart property is also responsible for toggling the robot's variable solverActive based on the selected part, set to true when the Handle is selected, and otherwise set to false. Lastly, we define translation and rotation speeds, in the moveSpeed and rotationSpeed variables, respectively, which provide some control over how quickly the selected part updates.

Once a part is selected, it will be transformed or rotated depending on the type of part selected and CurrentDirection. We will be using the two helper methods we defined in the base class that will be used for translating the direction into a vector we can pass to RobotController. This transformation is applied in the Update method; make this change now by amending the Update method with the following code:

```
    void Update()
    {
    if(CurrentPart == null)
    {
    return;
    }

    if(PlayStateManager.Instance.CurrentInteractible != null)
    {
    CurrentPart = null;
    return;
    }

    if (CurrentPart.Equals(PART_HANDLE))
    {
    PlayStateManager.Instance.Robot.MoveIKHandle(GetTranslationVector(moveSpeed
    * Time.deltaTime));
```

```
    }
    else
    {
    PlayStateManager.Instance.Robot.Rotate(CurrentPart,
    GetRotationVector(rotationSpeed * Time.deltaTime));
    }
    }
```

If we have a current part, we first determine the type of transformation required by the name, and, based on the name, either call the MoveIKHandle method using the GetTranslationVector we defined, or we call the Rotate method of RobotController, passing it the result of the GetRotationVector we defined earlier. Most of this code should look familiar to you as we are using the methods we had earlier used to control the robot with gestures.

The last task is to handle each distinct phrase recognized. We use a Dictionary to act as a delegate lookup, using the phrase as the key and delegate as the value. Make the following amendments in the code snippet:

```
delegate void KeywordAction(PhraseRecognizedEventArgs args);
Dictionary<string, KeywordAction> keywordCollection;

public override void StartHandler()
{
keywordCollection = new Dictionary<string, KeywordAction>();

keywordCollection.Add("rotate left", StartRotatingLeft);
keywordCollection.Add("rotate right", StartRotatingRight);

keywordCollection.Add("rotate up", StartRotatingUp);
keywordCollection.Add("rotate down", StartRotatingDown);

keywordCollection.Add("rotate one up", StartRotatingArm1Up);
keywordCollection.Add("rotate one down", StartRotatingArm1Down);

keywordCollection.Add("rotate two up", StartRotatingArm2Up);
keywordCollection.Add("rotate two down", StartRotatingArm2Down);

keywordCollection.Add("move up", StartMovingIKHandle);
keywordCollection.Add("move down", StartMovingIKHandle);
keywordCollection.Add("move left", StartMovingIKHandle);
keywordCollection.Add("move right", StartMovingIKHandle);
keywordCollection.Add("move forward", StartMovingIKHandle);
keywordCollection.Add("move back", StartMovingIKHandle);

keywordCollection.Add("stop", Stop);
```

```
    keywordRecognizer = new
    KeywordRecognizer(keywordCollection.Keys.ToArray(), confidenceLevel);
    keywordRecognizer.OnPhraseRecognized +=
    KeywordRecognizer_OnPhraseRecognized;
    keywordRecognizer.Start();
    }
```

Next, we have to make the delegation; we perform this in
the KeywordRecognizer_OnPhraseRecognized method using the recognized phrase as
the key to the lookup we just created and by invoking the associated delegate. Make the
following amendments to the KeywordRecognizer_OnPhraseRecognized method:

```
    void KeywordRecognizer_OnPhraseRecognized(PhraseRecognizedEventArgs args)
    {
    KeywordAction keywordAction;

    if (keywordCollection.TryGetValue(args.text, out keywordAction))
    {
    keywordAction.Invoke(args);
    }
    }
```

With this implemented, we now we have the laborious task of writing out all the delegates:

```
    void StartRotatingLeft(PhraseRecognizedEventArgs args)
    {
    CurrentPart = PART_BASE;
    CurrentDirection = Direction.Left;
    }

    void StartRotatingRight(PhraseRecognizedEventArgs args)
    {
    CurrentPart = PART_BASE;
    CurrentDirection = Direction.Right;
    }

    void StartRotatingArm1Up(PhraseRecognizedEventArgs args)
    {
    CurrentPart = PART_ARM_1;
    CurrentDirection = Direction.Up;
    }

    void StartRotatingArm1Down(PhraseRecognizedEventArgs args)
    {
    CurrentPart = PART_ARM_1;
    CurrentDirection = Direction.Down;
    }
```

```
    void StartRotatingArm2Up(PhraseRecognizedEventArgs args)
    {
CurrentPart = PART_ARM_2;
CurrentDirection = Direction.Up;
    }

    void StartRotatingArm2Down(PhraseRecognizedEventArgs args)
    {
CurrentPart = PART_ARM_2;
CurrentDirection = Direction.Down;
    }
```

There's little ambiguity for the phrases handled. Now, our next set of code is to handle phrases where the part is not obvious; to infer what they mean, we will make use of the user's gaze, specifically if the user is gazing at either arm 1 or arm 2, and ignoring the request if the user is gazing at neither:

```
    void StartRotatingUp(PhraseRecognizedEventArgs args)
    {
    if (!GazeManager.Instance.Hit)
    {
    return;
    }

    if(GazeManager.Instance.FocusedObject.name.Equals("arm 1",
StringComparison.OrdinalIgnoreCase))
    {
CurrentPart = PART_ARM_1;
CurrentDirection = Direction.Up;
    }
    else if (GazeManager.Instance.FocusedObject.name.Equals("arm 1",
StringComparison.OrdinalIgnoreCase))
    {
CurrentPart = PART_ARM_2;
CurrentDirection = Direction.Up;
    }
    }

    void StartRotatingDown(PhraseRecognizedEventArgs args)
    {
    if (!GazeManager.Instance.Hit)
    {
    return;
    }

    if (GazeManager.Instance.FocusedObject.name.Equals("arm 1",
StringComparison.OrdinalIgnoreCase))
    {
```

```
CurrentPart = PART_ARM_1;
CurrentDirection = Direction.Down;
}
else if (GazeManager.Instance.FocusedObject.name.Equals("arm 1",
StringComparison.OrdinalIgnoreCase))
{
CurrentPart = PART_ARM_2;
CurrentDirection = Direction.Down;
}
}
```

Our last method handles the move X keyword, where x is a direction. This method differs from the last one in that we are handling multiple keywords with a single delegate; we will use a regular expression to extract the direction:

```
void StartMovingIKHandle(PhraseRecognizedEventArgs args)
{
if(Regex.IsMatch(args.text, @"b(up|higher)b"))
{
CurrentPart = PART_HANDLE;
CurrentDirection = Direction.Up;
}
else if (Regex.IsMatch(args.text, @"b(down|lower)b"))
{
CurrentPart = PART_HANDLE;
CurrentDirection = Direction.Down;
}
else if (Regex.IsMatch(args.text, @"b(forward|away)b"))
{
CurrentPart = PART_HANDLE;
CurrentDirection = Direction.Forward;
}
else if (Regex.IsMatch(args.text, @"b(back|backwards)b"))
{
CurrentPart = PART_HANDLE;
CurrentDirection = Direction.Back;
}
else if (Regex.IsMatch(args.text, @"b(left)b"))
{
CurrentPart = PART_HANDLE;
CurrentDirection = Direction.Left;
}
else if (Regex.IsMatch(args.text, @"b(right)b"))
{
CurrentPart = PART_HANDLE;
CurrentDirection = Direction.Right;
}
}
```

With this method now implemented, we have finished the code for our first voice handler. The final task remaining is to wire it up in the editor, so head back to the Unity Editor and expand the Managers GameObject in the **Hierarchy** panel (if not already done), add a new empty GameObject with the name PSKeywordHandler by clicking on the **Create** dropdown, selecting **Create Empty**, and entering the appropriate name. Next, we need to add our script; select the newly created PSKeywordHandler **GameObject** and click on the **Add Component** button within the **Inspector** panel, type the name PSKeywordHandler, and select it when it becomes visible. Once attached, select the Managers GameObject and assign the GameObject PSKeywordHandler to the PlayStateManager script by clicking on and dragging the PSKeywordHandler onto the **Voice Handler** field.

Now we are good to take it for a test run; build and deploy the application to the device and return here once you have finished where we will move on to looking at how we can handle more complex (and natural) phrases.

Grammar Recognizer using Speech Recognition Grammar Specification (SRGS)

In the preceding section, we implemented the functionality to allow the user to control the robot using single words and short phrases, but this approach lacks the expressiveness that makes language so powerful and natural. In this section, we will explore an alternative approach that is more expressive and more inline with how people talk. We will be using GrammarRecognizer, which is available in the UnityEngine.Windows.Speech namespace, and will start off by looking at how to build a corpus of phrases we want to recognize, and then create a new PlayStateVoiceHandler to integrate it into the example.

The corpus of phrases we want to recognize will be written in an XML document that conforms to the SRGS, a standard governed by **World Wide Web Consortium** (**W3C**) specifically for defining a syntax used for a speech recognizer.

W3C is an international community made up of many organisations and members and is tasked with governing the World Wide Web. Governance is achieved through setting a set of standards to ensure openness. One standard, as mentioned earlier, is the SRGS, which is a standard that defines an XML schema used to define syntax for speech application. Microsoft supports this standard via their Speech API. You can learn more by visiting the official site at `https://www.w3.org/TR/speech-grammar/`.

Start by creating a new folder in your project's **Assets/App** directory called **StreamingAssets**. Unlike other files in your project, files that reside in the **StreamingAssets** folder of a Unity project are copied across to the filesystem of the destination platform. Once created, create an XML document (using an XML editor of your choice) called `srgs_robotcommands.xml`; this is the file that will contain our grammar.

The SRGS is a comprehensive schema, and I encourage you to investigate further if, you are curious about this subject; in this example, we will only be scratching the surface, but enough to put it to use in your own applications and, hopefully, enough to make you curious to learn more.

The general idea of the SRGS is to describe a set of phrases you are expecting from the user; its flexibility lies in how these phrases are defined. Unlike in the previous section where we were constrained to static words or short phrases, phrases in an SRGS are made up of a sequence of items, where items can be either single words, a sequence of words, or one of many alternatives. You also have the flexibility of making words optional, repetitive, and dynamically loaded at runtime. In the following diagram, we present a graphical representation of the phrases we are expecting from the user, something we will use as a reference when building up the SRGS:

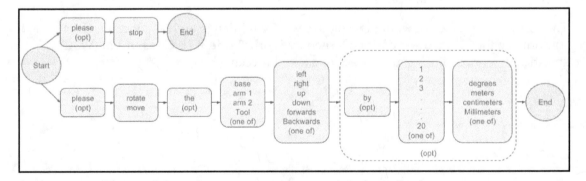

Here, the circles represent the entry and exit points, rounded rectangles represent items, and the arrows indicate the sequence. We have two paths, each for handling different intents, with the first (top) being used to recognize when the user wants to stop the current command and the second (bottom) showing the flow of a user committing a command. Here, we have items chained together, indicating whether an item is optional and/or a section of subitems; for example, the two phrases--please rotate the base left and rotate base left--provide the same meaning (or more correctly, satisfy the same flow).

For us, it's not enough just to recognize a phrase; ideally we want to be able to extract useful information from the user's utterance, for example, recognizing what part the user is wanting to move and how they want to move it. SRGS and GrammarRecgonizer provide us with this functionality and something we will make use of in this example, so let's jump into the document srgs_robotcommands.xml and start fleshing out the semantics.

We will start with our stop phrase, and despite its brevity, this short phrase will introduce everything we need to build out our longer, more complex phrase; with srgs_robotcommands.xml open, type or copy in the following:

```
<?xml version="1.0" encoding="UTF-8" ?>
<grammar version="1.0" xml:lang="en-US" mode="voice" root= "Entry"
xmlns="http://www.w3.org/2001/06/grammar" tag-format="semantics/1.0">

 <rule id="Entry" scope="public">
 <one-of>
 <item> <ruleref uri="#RobotStop"/> </item>
 </one-of>
 </rule>

 <rule id="RobotStop" scope="public">
 <example> please stop </example>
 <example> stop </example>

 <item repeat="0-1"> please </item>

 <one-of>
 <item> stop <tag> out.Action = "stop"; </tag> </item>
 </one-of>

 </rule>

</grammar>
```

As mentioned in the preceding code snippet, this short extract demonstrates everything we will need to build the rest of our grammar, so we will take our time on this first phrase and move quickly through the second.

The topmost root is the `grammar` tag. What is important here is the `root` tag; this value references the entry element in your grammar document, in our case, suitably named as `Entry`. This element is of the `rule` type, an element the engine uses to match the user's utterance with the defined phrases; it also provides a clean way of separating chunks of your phrase, thus allowing you to more easily reuse common expressions. Let's now define our entry point:

```
<rule id="Entry" scope="public">
<one-of>
<item> <ruleref uri="#RobotStop"/> </item>
</one-of>
</rule>
```

The contents of the `Entry` element are wrapped with a `one-of` element; this is akin to a switch statement in C#, allowing alternative expressions to be evaluated. Here, we only have one item, but we will return here in a bit to add reference to our second phrase. Next, we have the `item` element, which can contain utterances expected from the user and/or subelements, such as `ruleref`, `one-of`, or `tag`. For example, to recognize the `Hello Word` and `Hello Earth` phrases, we can define an `item` as follows:

```
<item> Hello <ruleref uri="#WorldPhrases" /> </item>
```

Here, as you might have suspected, `ruleref` points to another rule (either within the same file or externally) containing a set of alternatives for world (`World` and `Earth` in our case). Let's now inspect the referenced rule `RobotStop`:

```
<rule id="RobotStop" scope="public">
<example> please stop </example>
<example> stop </example>

<item repeat="0-1"> please </item>

<one-of>
<item> stop <tag> out.Action = "stop"; </tag> </item>
</one-of>

</rule>
```

In the preceding code snippet, we first provide two examples using the `example` element; like code comments, these are ignored by the recognizer, but are useful to us. Next, we define an optional utterance using the `repeat="0-1"` attribute and value, which means the recognizer will successfully match this phrase with or without the utterance `please` at the beginning of the user's utterance.

We have wrapped the next part in a `one-of` element with an item containing the utterance we are expecting--`stop`. For a successful match, the user's phrase must match this.

 You may have noted that we only have one element within one-of; in fact this element is not needed. The purpose of using it here is to introduce the element. As the name implies, it allows one of many items to be matched; for example, here we have only included the utterance `stop`, but we can easily add alternatives such as `halt`. To be successful, only one of these utterances will need to match.

As mentioned earlier, we need a way of extracting useful information, referred to as semantics, from the phrases recognized by the user, and this is where tag comes in; it provides a way of propagating values back up to the `GrammarRecognizer`. The contents of a tag can either be a value (`out = "stop"`) or key-value pair (`out.Action = "stop"`); here, we are using the latter, assigning `stop` to the `Action` key.

This now finishes our grammar definition for our first phrase--`stop`--and concludes our brief introduction to SRGS. We will now move on to fleshing out our second phrase, but will omit a lot of the details, as nothing new is introduced.

Start by extending our `Entry` element to include the reference to the second phrase; make the following amendments:

```
<rule id="Entry" scope="public">
<one-of>
<item> <ruleref uri="#RobotStop"/> </item>
<item> <ruleref uri="#RobotExecute"/> </item>
</one-of>
</rule>
```

We will continue drilling down, defining the top elements and then their dependencies. With this in mind, let's now define the `RobotExecute` rule:

```
<rule id="RobotExecute" scope="public">
<example> please rotate the base left </example>
<example> start rotating the base left </example>
<example> rotate the base left by 30 degrees </example>
```

```xml
<item repeat="0-1"> please </item>
<item repeat="0-1"> start </item>

<item>
<ruleref uri="#Action"/>
<tag> out.Action = rules.Action; </tag>
</item>

<item repeat="0-1"> the </item>

<item>
<ruleref uri="#Part"/>
<tag> out.Part = rules.Part; </tag>
</item>

<item>
<ruleref uri="#Direction"/>
<tag> out.Direction = rules.Direction; </tag>
</item>

<item repeat="0-1">
<item repeat="0-1"> by </item>

<item>
<ruleref uri="#Number"/>
<tag> out.Change = rules.Number; </tag>
</item>

<item>
<ruleref uri="#Unit"/>
<tag> out.Unit = rules.Unit; </tag>
</item>

</item>

<item repeat="0-1">
<one-of>
<item> please </item>
<item> thanks </item>
</one-of>
</item>

</rule>
```

Despite it being a fairly long piece of text, it introduces nothing new, apart perhaps from some nuances that will be described in the following section. If you compare it with the graphical representation we saw earlier, you will see how representative it is of the actual grammar we are defining.

The nuance I was referring to before is how we can propagate results back to the recognizer; let's take the `Part` item for example; the following is an extract taken from the preceding section:

```
<item>
<ruleref uri="#Part"/>
<tag> out.Part = rules.Part; </tag>
</item>
```

We wrapped the element `ruleref` around an `item` and included a `tag`. What might look out of place is the value assigned to `Part`, `rules.Part`; this is a syntax of SRGS that allows us to extract results from a rule we are referencing. To make it more concrete, let's examine the `Part` rule:

```
<rule id="Part">
<example> base </example>
<example> arm 1 </example>
<example> arm 2 </example>
<example> tool </example>

<one-of>
<item>
<ruleref uri="#PartBase"/>
<tag> out = "base"; </tag>
</item>
<item>
<ruleref uri="#PartArm1"/>
<tag> out = "arm 1"; </tag>
</item>
<item>
<ruleref uri="#PartArm2"/>
<tag> out = "arm 2"; </tag>
</item>
<item>
<ruleref uri="#PartTool"/>
<tag> out = "tool"; </tag>
</item>
</one-of>
</rule>
```

As can be seen from the preceding code snippet, the `Part` rule defines a set of alternatives, and, if matched, the item will assign a value to the out; for example, if `PartBase` is matched, then `base` is returned:

```
<item>
<ruleref uri="#PartBase"/>
<tag> out = "base"; </tag>
</item>
```

SRGS exposes the results from rules via the rules object, so, for this example, `rules.Part` will have the assigned results from the `Part` rule to our key `Part`. Let's continue drilling down our document, continuing with the individual part rules:

```
<rule id="PartBase">
<example> base </example>

<one-of>
<item> bottom </item>
</one-of>
</rule>

<rule id="PartArm1">
<example> arm 1 </example>

<one-of>
<item> arm 1 </item>
<item> arm one </item>
<item> lower arm </item>
<item> bottom arm </item>
</one-of>
</rule>

<rule id="PartArm2">
<example> arm 2 </example>

<one-of>
<item> arm 2 </item>
<item> arm two </item>
<item> upper arm </item>
<item> top arm </item>
</one-of>
</rule>

<rule id="PartTool">
<example> tool </example>

<one-of>
```

```
<item> tool </item>
<item> end </item>
<item> end effector </item>
<item> hand </item>
<item> handle </item>
</one-of>
</rule>
```

The preceding code snippet contains a set of rules to define a list of alternatives for each part. We could just as easily have embedded this into the `Part` rule, but just like code, this separation allows easier management and offers more resilience to change.

Let's now define our `Action` rule:

```
<rule id="Action">
<example> rotating </example>
<example> move </example>

<one-of>
<item> rotate <tag> out = "rotate"; </tag> </item>
<item> rotating <tag> out = "rotate"; </tag> </item>
<item> move <tag> out = "move"; </tag> </item>
<item> moving <tag> out = "move"; </tag> </item>
</one-of>
</rule>
```

Now, let's look at the `Directions` rule. As we are seeing a lot of repetition, we will omit a lot of the subrules. For the complete version, download the associated project:

```
<rule id="Direction">
<example> left </example>
<example> up </example>
<example> forward </example>

<one-of>
<item>
<ruleref uri="#DirLeft"/>
<tag> out = "left"; </tag>
</item>
<item>
<ruleref uri="#DirRight"/>
<tag> out = "right"; </tag>
</item>
<item>
<ruleref uri="#DirUp"/>
<tag> out = "up"; </tag>
</item>
<item>
```

```
<ruleref uri="#DirDown"/>
<tag> out = "down"; </tag>
</item>
<item>
<ruleref uri="#DirForwards"/>
<tag> out = "forwards"; </tag>
</item>
<item>
<ruleref uri="#DirBackwards"/>
<tag> out = "backwards"; </tag>
</item>
</one-of>
</rule>

<rule id="DirLeft">
<example> left </example>

<one-of>
<item> left </item>
<item> clockwise </item>
</one-of>
</rule>

...
```

Our final rules include `Number` and `Unit`; similar to the previous, we will omit a lot of the details, especially the `Number` rule:

```
<rule id="Unit">
<one-of>
<item> dgrees <tag>out = "degrees"; </tag> </item>
<item> meters <tag>out = "meters"; </tag> </item>
<item> centimeters <tag>out = "centimeters"; </tag> </item>
<item> millimetre <tag>out = "millimetre"; </tag> </item>
</one-of>
</rule>

<rule id="Number">
<one-of>
<item> zero <tag>out = 0; </tag> </item>
<item> one <tag>out = 1; </tag> </item>
<item> two <tag>out = 2; </tag> </item>
<item> three <tag>out = 3; </tag> </item>
<item> four <tag>out = 4; </tag> </item>
...
</one-of>
</rule>
```

This now completes our SRGS, but we have only just scratched the surface of what is possible. I encourage you to continue exploring and learning, especially as **Voice User Interfaces** (**VUIs**) have just started to be developed and, in no time, will become one of the dominant ways to interact with our digital peers. Furthermore and SRGS offers a flexible and comprehensive solution that can cater for a lot of use cases.

With our SRGS now defined, it's time to return to code and make use of it. As we did earlier, we will create a concrete implementation of our PlayStateVoiceHandler class, specifically to use our grammar file with GrammarRecognizer. With Unity Editor open, expand the App/Scripts folder in the **Project** panel and create a new script called PSSRGSGrammarHandler by clicking on the **Create** dropdown and selecting **C# Script**. Double-click on PSSRGSGrammarHandler to open it in Visual Studio.

Our newly created script PSSRGSGrammarHandler will resemble much of our PSKeywordHandler class, mainly because they are trying to achieve the same thing, with the main difference being how they interpret the recognized phrases from the user. Let's start by inheriting from the PlayStateVoiceHandler class and wiring up the GrammarRecognizer:

```csharp
using UnityEngine.Windows.Speech;
using System.IO;

public class PSSRGSGrammarHandler : PlayStateVoiceHandler
{
 public ConfidenceLevel confidenceLevel = ConfidenceLevel.Medium;

 public float rotationSpeed = 5.0f;

 public float moveSpeed = 5.0f;

 public override void StartHandler()
 {
 }

 public override void StopHandler()
 {

 }

 private void Update()
 {
 }

 private void OnDestroy()
 {
```

```
        }
    }
```

We will first include the `System.IO` and `UnityEngine.Windows.Speech` namespaces, the former to read our `srgs_robotcommands.xml` file, and the latter to get access to the `GrammarRecognizer`. We inherit from our `PlayStateVoiceHandler` class and implement the abstract methods. Finally, we will include a confidence threshold in the `conferenceLevel` variable, which we will use to filter out any results that fall below this level, and two variables exposing the speeds at which the rotate arm will rotate and move: `rotationSpeed` and `moveSpeed`. Next, we will flesh out the `StartHandler` and `StopHandler` abstract methods, which will be responsible for loading and disposing of the `GrammarRecognizer`. Make the following amendments to the `PSSRGSGrammarHandler` class:

```
public string SRGSFileName = "srgs_robotcommands.xml";

private GrammarRecognizer grammarRecognizer;

public override void StartHandler()
{
if(grammarRecognizer == null)
{
try
{
grammarRecognizer = new
GrammarRecognizer(Path.Combine(Application.streamingAssetsPath,
SRGSFileName));
grammarRecognizer.OnPhraseRecognized +=
GrammarRecognizer_OnPhraseRecognized;
}
catch
{
throw new Exception(string.Format("Error while trying to load or parse
the SRGS file {0}", SRGSFileName));
}
}

grammarRecognizer.Start();
}

public override void StopHandler()
{
if(grammarRecognizer != null)
{
grammarRecognizer.Stop();
}
```

```
    }

    private void Update()
    {
    }

    private void OnDestroy()
    {
    if (grammarRecognizer != null)
    {
    grammarRecognizer.Stop();
    grammarRecognizer.OnPhraseRecognized -=
    GrammarRecognizer_OnPhraseRecognized;
    grammarRecognizer.Dispose();
    grammarRecognizer = null;
    }
    }

    private void
    GrammarRecognizer_OnPhraseRecognized(PhraseRecognizedEventArgs args)
    {
    }
}
```

We add two new variables, one to store the location and filename of our SRGS file, and the other to hold reference to an instance of the GrammarRecognizer. In the StartHandler method, we instantiate an instance of the GrammarRecognizer, passing in the file path of our srgs_robotcommands.xml file. Next, we register our GrammarRecognizer_OnPhraseRecognized delegate, and then call Start on the GrammarRecognizer, which will, as you might expect, start our GrammarRecognizer.

Within the StopHandler method, we simply call Stop on the GrammarRecognizer and take care of unregistering the event handler and disposing of GrammarRecognizer in the OnDestory method. Finally, we add our GrammarRecognizer_OnPhraseRecognized handler.

With just the previously written code, we have a functional GrammarRecognizer. Our last task is to handle the recognized phrases, but, before we do, let's quickly discuss what the GrammarRecognizer returns.

KeywordRecognizer and GrammarRecognizer both inherit from PhraseRecognizer, and both return a PhraseRecognizedEventArgs instance when a phrase is recognized. While both share the type class, it's only the GrammarRecognizer that makes use of the semanticMeanings property (of the SemanticMeaning type). This type encapsulates the recognized semantics within your phrase; for example, in our SRGS, we define an output variable--Action--which is set to either "stop", "rotate", or "move"; if matched, these values are returned via the PhraseRecognizedEventArgs
using the semanticMeaning property. We can iterate through the returned semantics (semanticMeaning is an array) and match the key with a given variable we are expecting (in this case, Action) and then obtain its value (or values if there are multiple results).

 KeywordRecognizer and GrammarRecognizer cannot run at the same time; before using one, you must explicitly stop the other.

To mitigate the possibility of bugs, we will define some constants for the semantic keys we are expecting; make the following amendments to the PSSRGSGrammarHandler class:

```
sealed class SemanticKeys
{
public const string Action = "Action";
public const string Part = "Part";
public const string Direction = "Direction";
public const string Change = "Change";
public const string Unit = "Unit";
}

sealed class CommandAction
{
public const string Stop = "stop";
public const string Rotate = "rotate";
public const string Move = "move";
}

sealed class CommandUnit
{
public const string Degrees = "degrees";
public const string Meters = "meters";
public const string Centieters = "centieters";
public const string Millimeters = "millimeters";
}
```

By looking at the graphical representation of the phrases we are interested in. We can see an opportunity to encapsulate the details into some structure. For example, the phrase (or command) is either going to stop or manipulate a specific part of the robot. If the latter, then we are expecting reference to the part, direction, and possibly some discrete change. Encapsulating it into a structure keeps our code cleaner and better prepares the code base for future requirement changes. Let's now define a struct that encapsulates the parameters of our phrase; make the following amendments to the `PSSRGSGrammarHandler` class:

```
 public struct Command
 {
public string action;
public string part;
public string direction;
public string unit;
public float? change;

public bool IsDiscrete
{
get { return change.HasValue && (unit != null && unit != string.Empty); }
}

public float ScaledChange
{
get
{
if (!change.HasValue)
{
return 0;
}

return change.Value * GetMeterScaleForUnit(unit);
}
}

public float GetMeterScaleForUnit(string unit, float defaultScale=1f) {
if(unit == null || unit == string.Empty)
{
return defaultScale;
}

switch (unit)
{
case CommandUnit.Centieters:
return change.Value / 100f;
case CommandUnit.Millimeters:
return change.Value / 1000f;
}
```

```
    return defaultScale;
    }
}
```

In the preceding code snippet, we define a simple struct that encapsulates the possible values returned by the `GrammarRecognizer`; we have added two convenient methods: the `IsDiscrete` property that returns true if the user's intention is for discrete movement, and false otherwise, and `GetMeterScaleForUnit`, which is used to standardize the value requested by the user to meters (the unit we are working with in Unity). Before moving on to handle the response from the `GrammarRecognizer` and binding the return values with the `Command` struct, now let's define variable and property to hold reference to the current command, unsurprisingly named `CurrentCommand`. The reason we do this is, as we did in the preceding section, if `Command` is not discrete then it will be executed continuously until explicitly told to stop:

```
private Command? _currentCommand;

public Command? CurrentCommand
{
get
{
return _currentCommand;
}
private set
{
if (_currentCommand.HasValue)
{
if(_currentCommand.Value.part.Equals(PART_HANDLE,
StringComparison.OrdinalIgnoreCase))
{
PlayStateManager.Instance.Robot.solverActive = false;
}

_currentCommand = value;

if (_currentCommand.Value.part.Equals(PART_HANDLE,
StringComparison.OrdinalIgnoreCase))
{
PlayStateManager.Instance.Robot.solverActive = true;
}
}
}
}
```

Within the CurrentCommand property, we handle toggling the solveActive variable of RobotController every time a Command is set. This ensures that the robot is unlikely, to ever be in an invalid state with the inverse kinematics solver active when it shouldn't be.

We are almost there; the last two major tasks remaining are creating and binding a Command to a recognized phrase and then actually processing the CurrentCommand. Let's start with binding; this, of course, is performed when we are notified by the GrammarRecognizer of a valid match via the OnPhraseRecognized event. Make the following amendments to the GrammarRecognizer_OnPhraseRecognized method:

```
    private void
GrammarRecognizer_OnPhraseRecognized(PhraseRecognizedEventArgs args)
    {
    if(args.confidence < confidenceLevel)
    {
    return;
    }

    Command commandCandidate = CreateCommand(args);

    if (IsCommandValid(commandCandidate))
    {
    CurrentCommand = commandCandidate;
    }
    }
```

We first ensure that the interpreted phrase has reached our confidence threshold and, if satisfied, we delegate the creation and binding of the Command instance to the CreateCommand method passing over the received argument PhraseRecognizedEventArgs. The CreateCommand method is responsible for iterating through each semanticMeanings assigned to the parameter and binding them to the relevant variable of our newly created command object. Once this is returned, we verify that it is valid before assigning it to the property CurrentCommand (implemented earlier).

Let's now implement our creation and validation methods; add the following methods to the PSSRGSGrammarHandler class:

```
    Command CreateCommand(PhraseRecognizedEventArgs args)
    {
    SemanticMeaning[] meanings = args.semanticMeanings;

    return new Command
    {
    action = meanings.Contains(SemanticKeys.Action) ?
    meanings.SafeGet(SemanticKeys.Action).Value.values[0] : string.Empty,
```

```
  part = meanings.Contains(SemanticKeys.Part) ?
meanings.SafeGet(SemanticKeys.Part).Value.values[0] : string.Empty,
  direction = meanings.Contains(SemanticKeys.Direction) ?
meanings.SafeGet(SemanticKeys.Direction).Value.values[0] : string.Empty,
  change = meanings.Contains(SemanticKeys.Change) ?
int.Parse(meanings.SafeGet(SemanticKeys.Change).Value.values[0]) : 0,
  unit = meanings.Contains(SemanticKeys.Unit) ?
meanings.SafeGet(SemanticKeys.Unit).Value.values[0] : string.Empty
  };
  }


  bool IsCommandValid(Command command)
  {
   // details omitted for brevity
  }
```

The `CreateCommand` class simply returns a new value of the `Command` with its properties
bound to the available semantics of the argument `PhraseRecognizedEventArgs`. The next
method, `IsCommandValid`, is responsible of ensuring that the `Command` is in a valid state
for processing; this is a fairly verbose method and for this reason, has been omitted here,
but you can check it out in the full source available for download at this book's website.

 The `semanticMeanings` property of `PhraseRecognizedEventArgs` is
an array; to ensure that the code was compact enough to publish, the
`Contains` and `SafeGet` extension methods were created for convenience.
The implementation of these extensions can be found in the
`Extensions.cs` file accompanying with this project.

Now we can listen to and understand the user's utterance (that match our predefined
syntax). The last piece of code that needs to be written is to execute the command, most of
which should look familiar to you as it shares a lot with the code in the preceding section;
add the following method to your `PSSRGSGrammarHandler` class:

```
void ProcessCurrentCommand()
{
if (!CurrentCommand.HasValue)
{
return;
}

Command command = CurrentCommand.Value;

switch (command.action)
{
case CommandAction.Stop:
{
```

```
        // terminate command
        CurrentCommand = null;
        break;
      }
      case CommandAction.Rotate:
      {
        PlayStateManager.Instance.Robot.solverActive = false;

        if (command.IsDiscrete)
        {
          PlayStateManager.Instance.Robot.Rotate(command.part,
        GetRotationVector(command.direction, command.ScaledChange));
          CurrentCommand = null;
        }
        else
        {
          PlayStateManager.Instance.Robot.Rotate(command.part,
        GetRotationVector(command.direction, rotationSpeed * Time.deltaTime));
        }

        break;
      }
      case CommandAction.Move:
      {
        PlayStateManager.Instance.Robot.solverActive = true;

        if (command.IsDiscrete)
        {
        PlayStateManager.Instance.Robot.MoveIKHandle(GetTranslationVector(command.d
        irection, command.ScaledChange));
          PlayStateManager.Instance.Robot.solverActive = false;
          CurrentCommand = null;
        }
        else
        {
        PlayStateManager.Instance.Robot.MoveIKHandle(GetTranslationVector(command.d
        irection, moveSpeed * Time.deltaTime));
          PlayStateManager.Instance.Robot.Rotate(command.part,
        GetRotationVector(command.direction, rotationSpeed * Time.deltaTime));
        }
        break;
      }
    }
  }
```

We control which block is executed based on the bound `action` assigned to the command, and for each action, we first check whether the command is discrete or not. If discrete, we execute it using the associated direction and offset before terminating the command by setting it to `null`. If not discrete, we adjust the part using the direction and related speed (`rotationSpeed` or `moveSpeed`). To execute commands continuously, we need to call this method's each frame; we can easily do this by adding a call in the `Update` method. Let's add that now; make the following amendments to the `Update` method:

```
private void Update()
{
if(CurrentCommand.HasValue)
{
ProcessCurrentCommand();
}
}
```

This now completes our `PSSRGSGrammarHandler` class; the only thing left to do is to hook this class up in the editor and test it. Jump back into the Unity Editor, and, as we did with the `PSKeywordHandler`, expand the `Managers` GameObject in the **Hierarchy** panel (if not already done), add a new empty GameObject with the name `PSSRGSGrammarHandler` by clicking on the **Create** dropdown, selecting **Create Empty**, and entering the appropriate name. Next, we need to add our script--select the newly created `PSSRGSGrammarHandler` GameObject and click on the **Add Component** button within the **Inspector** panel, typing and selecting `PSSRGSGrammarHandler` . Once attached, select the `Managers` GameObject and assign the `PSSRGSGrammarHandler` to the `PlayStateManager` script by clicking on and dragging the `PSSRGSGrammarHandler` onto the **Voice Handler** field. The **Inspector** panel of the `Managers` GameObject should look similar to this:

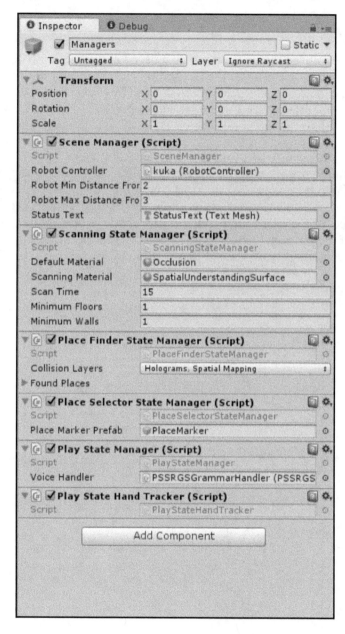

Try out the new voice handler by building and deploying to your device.

Summary

In this chapter, we explored the ways in which a user can interact with holograms; we started by looking at how we can use the navigation and manipulation gestures to control the robot arm, and extended this further by looking at how we can allow the user to actually *touch* the hologram. We did this by tracking the user's hand and responding to collisions with the hologram.

In the last part, we explored two approaches to integrating voice into your application, one that dealt with simple phrases, and the other using SRGS to handle more complex (and natural) phrases. We have covered a lot in this chapter; focusing on two interaction paradigms, gestures and voice, and explored how we might use them to deliver a more intuitive experience for the user to interact with the virtual world. In the next chapter, we will explore one of the pinnacle use cases of HoloLens--collaborative workflows. In other words we will be, looking at how we can share the same content between two users.

7
Collaboration with HoloLens Using Unity

In this chapter, we further explore what it means to bring computing into the physical world, specifically looking at the opportunities for improving collaboration by unlocking digital content from behind the screen and placing it in a communal space. Along with this, we will also explore how we can use physical objects to interact with the digital world. Let's unpack each of these, starting with how MR can improve collaboration.

I would argue that screen-based consumption creates a silo experience, as it draws all of our attention away from everything and everybody else. This is true in the context of work, such as working on our computers, as well as many forms of entertainment. Of course, there are times this makes sense (coding being one that comes to mind) but there are plenty of other times when digital content should be shared and discussed collaboratively. A good example illustrating this is a group of friends watching a sports game; the current medium (the screen) requires each friend to sit side by side with his/her attention focused on the screen; their posture and focus discouraging any form of communication. This is nicely illustrated by one of Microsoft's concepts where an NFL game is visualized onto a tabletop in the center of the room; you can see this video here `https://youtu.be/HvYj3_VmW6I?t=1m8s`.

Another, more relevant to this chapters example, concept is shown in the following images, illustrating how designers can use the HoloLens to project their work into the real-world, giving an more accurate sense of scale and better opportunities to collaborate.

Microsoft HoloLens and CreativeA Design, Source: Image courtesy of Microsoft Corporation, Source: https://www.microsoft.com/en-us/hololens

Being able to bring the digital content into the real world allows peers to position themselves around a table, facing each other as you would during dinner; at a cafe, or around a camp fire. This organization of people naturally encourages social interaction, for good or bad. There are plenty of examples where this is applicable, in both a professional and personal capacity and this is really the core of this chapter.

Our next main topic centers around extending the interface into the real world; the concept of manipulating digital content using something physical (or tangible) is known as **Tangible User Interfaces** (**TUIs**), in part conceived by Hiroshi Ishii, a professor in the MIT Media Laboratory who heads the Tangible Media Group. TUIs (or, more broadly, Tangible Interaction) is concerned with interfaces that employ physical objects, surfaces, and spaces as tangible embodiments of digital information and processes. Consequently, the motivation for this was the dissatisfaction with screen-based interfaces (including VR), as they are seen as estranging people from *the real world* (seeing a trend here). HoloLens, and MR in general, provides us with the opportunity to not only integrate digital content into the physical world but also integrate the physical world with the digital.

In this chapter's example, we will explore these concepts by developing a hypothetical application for 3D designers; our task is to allow designers to pull their designs out from the screen and into the real world for better collaboration, to inspect and review their work with peers. We will further decouple the computer from the content by extending some controls into the physical world. By the end of this chapter, you will have learned the following:

- Export, share, and import `WorldAnchor`, allowing spaces to be shared across devices
- Manipulate digital content using physical objects using the Vuforia plugin
- Use spatial sound to assist the user in locating the hologram

Time to get started; we will start our journey with a quick explanation of the example we will be building upon before looking at integrating Vuforia, then sharing, and finally, adding spatial sound.

Project setup

In this section, we will flesh out the details of the project and walk through the details of the starter project and accompanying components before jumping into extending it for collaborative review and augmenting physical objects to manipulate the hologram. If you haven't already, clone or download the repository for this book from `https://github.com/PacktPublishing/Microsoft-HoloLens-By-Example`. The project we will be working on is located in the `Starter` folder found in the directory `Chapter7/Starter`. In addition, there is a Blender plugin in the parent directory called `BlenderLIVESrv.zip`, which we will walk through installing shortly, but before that, let's continue on with our fictional project.

As mentioned earlier, our task is to develop a collaborative design tool for 3D designers. We will achieve this by allowing 3D designers to share their content from within Blender and, once shared, the model will be made available to nearby HoloLens devices. If the model has not already been placed, the user will be able to place it on a surface nearby (something we have had plenty of practice with). Once placed, other HoloLens users will be able to join in with the session and preview the content from where it has been placed. To assist the user in locating the model, we will use spatial sound. Finally, to allow collaborative interaction, we will expose a way for the user to rotate and translate the model using physical objects.

Blender is a free and open source 3D computer graphics software product used for creating 3D content for anything ranging from animated films, visual effects, art, 3D printed models, to interactive 3D applications, and video games. To follow along with this chapter, you will need to download and install it. The installation file and instructions (minimal) can be found at `https://www.blender.org`.

Installing and using the Blender plugin

With our task fairly well defined, let's briefly discuss how the Blender plugin works, including how to install and use it. Blender makes itself extendable via plugins. After installing the accompanying plugin, BlenderLIVE, the user will be able to share selected models. As soon as one object is shared, the service will start broadcasting itself across the local network. It is this broadcast that the HoloLens listens out for and, once discovered, will use to establish a connection with the service (BlenderLIVE). Once a connection is made, the associated data is shared with the HoloLens, including the model's geometry, materials, and textures.

Once all of the data has come across, the user will be asked to place the object onto a surface in their environment, at which point, we can anchor it and share this anchor with other users, thus, creating a shared and collaborative space.

Any user will be able to interact with the model using the physical objects. These interactions are serialized into operations and sent to the BlenderLIVE service, where it propagates the update to the local model (within the Blender editor) and to connected peers. The following figure provides a pictorial description of what we have just discussed:

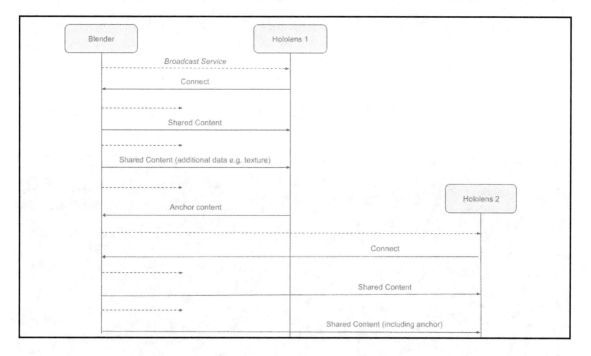

Now, with an understanding of what the Blender plugin is responsible for, let's walk through the steps required to install and use it. If you have not done so already, head over to `https://www.blender.org` to download and install Blender using the instructions provided by the installer. Once installed, launch Blender and follow the following steps to install the plugin:

1. Open the User Preferences dialog via the menu **File** | **User Preferences**.
2. Select the tab **Add-ons** from the **Blender User Preferences** dialog.
3. Click **Install** from the file button located at the bottom of the dialog.

4. Select the `BlenderLIVESrv.zip` file located in the `CollaborativeDesign` folder of this book's repository.

5. Select **Object** from the **Categories** panel.

6. Locate and check **Object: BlenderLIVE Service**.

The following screenshot shows an example of the **Blender User Preferences/Add-on** dialog:

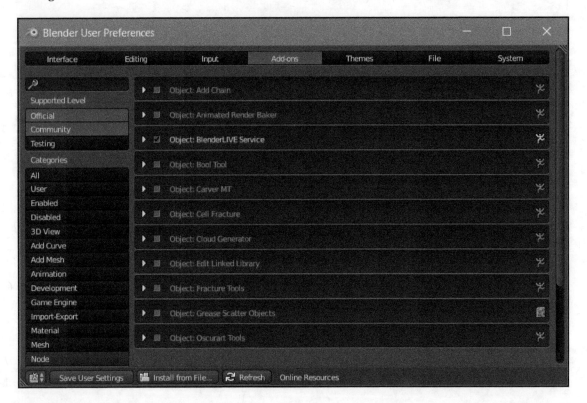

To share an object, close the **Blender User Preferences** dialog and select an object you would like to share by right-clicking on it with using your mouse. Now, with the object selected, select the **Object** tab from the tool shelf located bottom right of the Blender workspace (shown in the following screenshots):

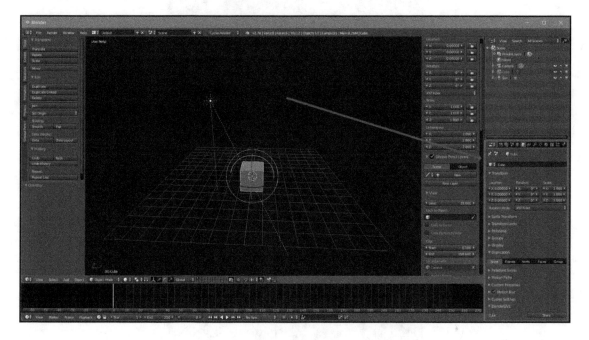

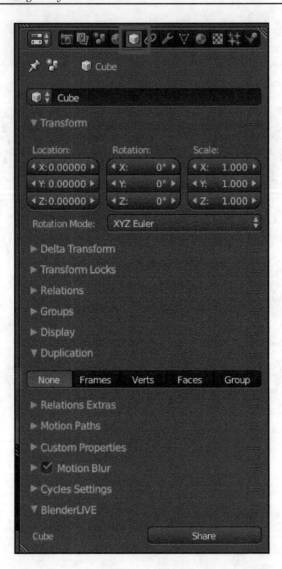

From here, it's simply a matter of expanding the **BlenderLive** panel and clicking on the **Share** button to activate sharing for this model to any listening device; as shown in the following screenshot:

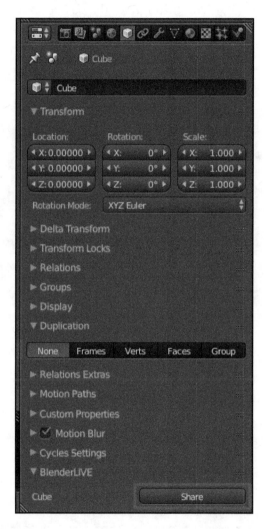

With that, we have finished our introduction to the Blender plugin in this project. Next, we will take a quick look at the existing `Starter` project.

This project is very much a proof of concept and developed specifically for this book to provide an interesting example to work with. For this reason, I recommend using basic models with a relatively small set of vertices, materials, and textures.

Unity starter project

When you open the `Starter` project in Unity, you will be greeted with a familiar sight. It's the setup we have been working with throughout the past few chapters. Here, we will briefly describe each major component before looking at how we can download and view the shared object from Blender:

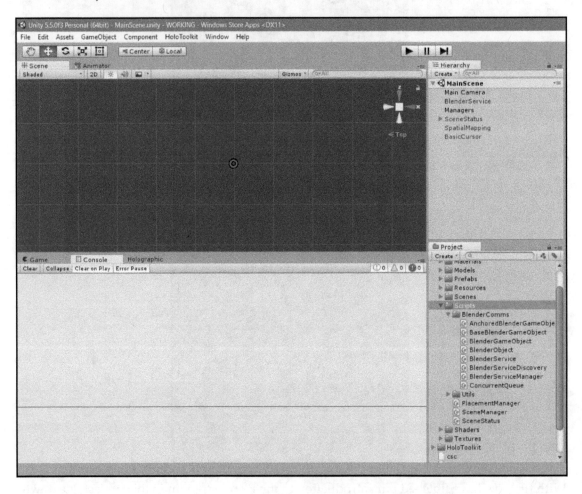

As we have done previously, we are building upon the work we have already covered to allow us to solely focus on the main topics of this chapter, namely, `WorldAnchor`, integrating Vuforia, and spatial sound. The main difference is the GameObject **BlenderService** from within **Project Hierarchy**.

This has the components (scripts) `BlenderServiceDiscovery`, `BlenderService`, and `BlenderServiceManager` attached; these are responsible for handling the discovery (`BlenderServiceDiscovery`) and communication with the BlenderLIVE service introduced earlier (`BlenderService`), thus, acting as a interface between the Blender service and Unity (`BlenderServiceManager`). In terms of actual coding, we will be amending the `AnchoredBlenderGameObject` script and `SceneManager` along with adding new scripts as and when required. With that being said, this concludes our quick introduction to the starter project. Time to code.

 This examples requires a connection between the BlenderLIVE service and each HoloLens. It uses port 8881 to broadcast the service across **User Datagram Protocol (UDP)** and port 8080 to establish a connection **Transmission Control Protocol (TCP)** with each client. Please ensure that these ports are open on your host computer (you may require some updates to your firewall).

The project

At this stage, you should have a good understanding of what it is we are going to build and a general idea of how it is currently strung together. In this section, we will turn this concept into a working prototype, initially starting with wiring up the BlenderLIVE service, admittedly not specific to HoloLens. But it will give you a better understanding of the application. We will then integrate and make use of Vuforia, a computer vision tool specifically for image recognition and tracking. After this, we will look at how to create shared spaces between devices, finally wrapping up by adding spatial sound to help users locate the hologram when placed. Let's get started.

Connecting to BlenderLIVE

We will communicate with the BlenderLIVE service using the `BlenderServiceManager` class via the `SceneManager` class. So, jump into the `SceneManager` script, which you can quickly locate by entering `SceneManager` into the search field of the **Project** panel. Once it's visible, double-click it to open it in Visual Studio.

Once opened, navigate to the `Start` method and make the following amendments:

```
void Start ()
{
  BlenderServiceManager.Instance.OnBlenderServiceStateChanged +=
  BSM_OnBlenderServiceStateChanged;
  BlenderServiceManager.Instance.OnBlenderGameObjectCreated +=
  BSM_OnBlenderGameObjectCreated;
  BlenderServiceManager.Instance.OnBlenderGameObjectUpdated +=
  BSM_OnBlenderGameObjectUpdated;
  BlenderServiceManager.Instance.OnBlenderGameObjectDestoryed +=
  BSM_OnBlenderGameObjectDestoryed;

  SceneStatus.Instance.SetText("Scanning room", 3.0f);
}
```

Because of the asynchronous nature of the service, we will be interacting with the `BlenderServiceManager` class via events. The following list briefly describes each event that we have subscribed before:

- `BSM_OnBlenderServiceStateChanged` is called when the state of the connection changes. We can use this to monitor when the service is discovered, connected, and disconnected.
- `BSM_OnBlenderGameObjectCreated` is called when a shared model is first created, `BSM_OnBlenderGameObjectUpdated` is called when the model is updated, and `BSM_OnBlenderGameObjectDestoryed` is called when unshared within Blender.

The last statement, `SceneStatus.Instance.SetText("Scanning room", 3.0f)`, is used to set our tag-along status label, which we are using to keep the user informed of the application's state. Let's now stub out those delegates to remove the errors. Add the following code to the `SceneManager` class:

```
private void BSM_OnBlenderGameObjectDestoryed(string name) { }

private void BSM_OnBlenderGameObjectCreated(BaseBlenderGameObject bgo)
{ }

private void BSM_OnBlenderGameObjectUpdated(BaseBlenderGameObject bgo)
{ }

private void
BSM_OnBlenderServiceStateChanged(BlenderServiceManager.ServiceStates
state) { }
```

We now have the methods the `BlenderServiceManager` will use to communicate with us. All we need to do now is ask it to start, but before we do just that, let's quickly review what starting this service will actually do.

As mentioned earlier, the `BlenderServiceManager` interfaces between `BlenderService` and Unity. Its first task is to discover the service (if active on the local network). Discovery is achieved by listening to a UDP emitted by the BlenderLIVE service every once in a while. Once it is discovered, `BlenderServiceManager` will automatically connect by establishing a TCP connection and proceed to download any shared models. `BlenderServiceManager` also takes care of constructing the shared GameObjects and making the application aware via the event `BSM_OnBlenderGameObjectCreated` (when created). After this, it will be the responsibility of the application to place and display the model. But before starting the service, it makes sense that we have scanned enough of the room that there is sufficient surface to place the model onto when it has been downloaded. More importantly, when downloading with an accompanying `WorldAnchor`, it's important that the scanning has been sufficiently comprehensive such that the HoloLens has enough understanding of the environment to be able to recognize where the shared `WorldAnchor` belongs. But that's getting a little ahead of ourselves. More on this later.

 As discussed in the previous chapters, `WorldAnchor` describes a physical position and rotation in the real world and how it relates to the position and rotation of the application's frame of reference. In Unity, attaching a `WorldAnchor` locks a GameObject's position to the physical space.

As we have done in the previous chapters, we will take the simplest approach and assume the user has sufficiently scanned the area after a certain time and, once this time has lapsed, we will ask `BlenderServiceManager` to discover and connect to the BlenderLIVE service. Head back to the top of the class and add the following variable to determine this time threshold:

```
public float initialScanningTime = 10.0f;
```

The most convenient place to monitor whether this time has elapsed is in the `Update` method, so head back down and make the following amendments:

```
void Update ()
{
  if (BlenderServiceManager.Instance.ServiceState ==
  BlenderServiceManager.ServiceStates.Stopped)
  {
    if (Time.time - SpatialMappingManager.Instance.StartTime >
    initialScanningTime)
    {
      SceneStatus.Instance.SetText("Searching for BlenderLIVE Service");
```

```
        BlenderServiceManager.Instance.SearchAndConnectToService();
      }
    }
  }
```

Here, we are testing the state of `BlenderServiceManager`. If stopped and sufficient time has lapsed, we will call `SearchAndConnectToService` which, as the name suggests, listens out for the service and, once found, will establish a connection.

As mentioned previously, `BlenderServiceManager` communicates via events. One such event notifies us of the state of the service. Let's fill this in; here, we will just pass those state changes to the user, keeping them informed of the state of the application. Scroll down to the method `BSM_OnBlenderServiceStateChanged` and make the following amendments:

```
private void
BlenderServiceManager_OnBlenderServiceStateChanged(BlenderServiceManager.Se
rviceStates state)
  {
    switch (state)
    {
      case BlenderServiceManager.ServiceStates.Discovered:
      SceneStatus.Instance.SetText("Found BlenderLIVE service", SHORT);
      break;
      case BlenderServiceManager.ServiceStates.Connecting:
      SceneStatus.Instance.SetText("Connecting to BlenderLIVE service",
      SHORT);
      break;
      case BlenderServiceManager.ServiceStates.Connected:
      SceneStatus.Instance.SetText("Connected to BlenderLIVE service",
      SHORT);
      break;
      case BlenderServiceManager.ServiceStates.Disconnected:
      SceneStatus.Instance.SetText("Disconnected from BlenderLIVE
      service", LONG);
      break;
      case BlenderServiceManager.ServiceStates.Failed:
      SceneStatus.Instance.SetText("BlenderLIVE connection failednPlease
      restart everything and try again.", LONG);
      break;
    }
  }
```

Here, we are simply updating the status text with the current state of the service. Similarly, we want to notify the user when an object has been unshared; find the method `BSM_OnBlenderGameObjectDestoryed` and make the following amendments:

```
private void BlenderServiceManager_OnBlenderGameObjectDestoryed(string
name)
{
SceneStatus.Instance.SetText(string.Format("{0} removed", name), MEDIUM);
}
```

The last two delegates remaining are concerned with when an object is created (first retrieved) and when the object is updated. When an object is created, we will pass it onto `PlacementManager`, whose responsibility it is to assist the user in placing the object and notified via the `OnObjectPlaced` event once the object has been placed. By contrast, when the object is updated, we will simply clear the status text.

Let's make those amendments to the `BSM_OnBlenderGameObjectCreated` and `BSM_OnBlenderGameObjectUpdated` methods:

```
private void BSM_OnBlenderGameObjectUpdated(BaseBlenderGameObject bgo)
{
SceneStatus.Instance.SetText("");
}

private void BSM_OnBlenderGameObjectCreated(BaseBlenderGameObject bgo)
{
SceneStatus.Instance.SetText("Blender object ready for
placementnAir-tap on a suitable surface to place", MEDIUM);
PlacementManager.Instance.AddObjectForPlacement(bgo);
}
```

Our final task is to listen out for when the object has been placed; head back to the `Start` method and make the following amendment:

```
...
PlacementManager.Instance.OnObjectPlaced +=
PlacementManager_OnObjectPlaced;
SceneStatus.Instance.SetText("Scanning room", 3.0f);
```

And now add in the delegate method:

```
private void PlacementManager_OnObjectPlaced(BaseBlenderGameObject
bgo)
{
    SceneStatus.Instance.SetText("");
}
```

With that implemented, we can pull models from Blender into the real world. Now is a good time to build and deploy, testing whether everything is working before we continue.

Remember that you will need to have Blender and the BlenderLIVE service running with at least one object shared. See the preceding section for details. If you do run into trouble, check that you have opened the ports discussed earlier.

The following image shows a live link between Blender and HoloLens:

Understanding and using the real world

One of the opportunities (and challenges) with MR is understanding enough of the real world to make it useful in the experiences you create. We have seen how we can make use of our understanding of the surrounding surfaces, but imagine how much more magical the experience would be if we could recognize objects and make appropriate use of them. For example, being able to recognize a chair would give us the ability to let virtual characters sit on the chair or know that a holographic phone (the ones that sit on a desk, if they still exist) is best placed on a desk.

This process of recognizing objects from a digital camera feed sits in the field of computer vision. Luckily, for us, there are a few libraries out there that bundle together a suite of tools to help us better understand the image; one of those libraries is Vuforia. A computer vision library born from the research labs at Qualcomm, it quickly became the de facto AR tool amongst many budding mobile application developers during the early 2000s. Now, owned by PTC, Vuforia still remains popular and provides us with a straightforward way of implementing our proof of concept of being able to manipulate our hologram using physical objects, which is the topic of this section.

To use Vuforia, you will first need to sign up as a developer. It's free (as far as I'm aware), so head to `https://developer.vuforia.com/` and register (or login if you already have an account). To make use of Vuforia, we need to obtain a license key and register/create targets for our application. Licensing is simply a matter of registering your application and its purpose. Let's do that now. Once logged in, head to `https://developer.vuforia.com/targetmanager/licenseManager/licenseListing` and click on the **Add License Key** button. Next, select the **Development - my app is in development** option along with entering a compelling name for your application. Once you click **Next** and agree to the terms and conditions, you will be taken back to the **Licence Manager** screen with, assuming everything went smoothly, your application on a list. Click on the name to see the details, including the licence, and make note of this as you will need it shortly.

 As mentioned before, Vuforia is a library that provides a way to recognize and track targets. Targets can be anything from a planar (flat) image, a cuboid (box), cylinder, or 3D object, and is normally predefined. To register our targets, we will use Vuforia's online **Target Manager** tab, which, at the time of writing, is visible next to **Licence Manager**. Click to open the relevant page:

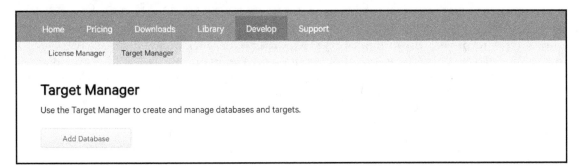

Next, we need to create a Database, which is essentially a repository for a set of targets; create this database by clicking on the **Add Database** button, as shown. This will open a dialog prompting you to enter the details for this database. Enter a name and leave **Type** as **Device**:

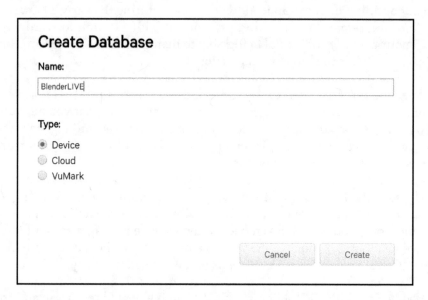

Types refer to the types of target used and where they will reside; **Device** and **Cloud** are self-explanatory, while **VuMark** is Vuforia's proprietary bar code. For more information about target types, refer to the documentation at, `https://library.vuforia.com/ articles/Training/Getting-Started-with-the-Vuforia-Target-Manager`.

After clicking **Create**, you will be able to access the newly created database. The last few steps add our targets and then download the convenient Unity package, which we will import into our project. Back on the **Target Manager** page, click on the database name you just created. To create a new target, click on the **Add Target** button. This will open a dialog, where you define your target. As mentioned before, there are a few types of target and you can learn all about them from the link shown arlier. For us, we simply want **Single Image**, so leave **Type** as its default.

Next, we need to provide it with an image. There are many approaches to recognizing objects from a camera frame. Vuforia favours images with a lot of features. Here, features are distinct corners. You can use any image or use one of the images I have included in the **Texture** folder of the project. Next, we need to provide the **Width**; the target will be in augmented reality. For this project, I have chosen 0.066 meters (or 6.6 centimeters). Height can be ignored, as it is inferred by the aspect ratio of your image and the given width. The last piece of data required is the target's **Name**. Enter a meaningful name in the **Name** field shown in the following screenshot:

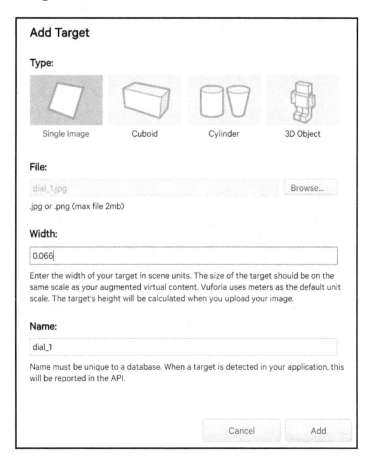

Once clicking on **Add**, you will be returned to the database's Target page with your newly created target visible. Along with its **Name, Type, Status**, and **Date Modified**, it includes **Status**. This status provides some indication of how confidently Vuforia is able to recognize the target. We will be using one target for rotation and another for translation, so go ahead and repeat the process to add one more targets (a different image).

The final task remaining is to download the package. To do this, simply click on the **Download Database (All)** button and select **Unity Editor** as the development platform as in the following screenshot:

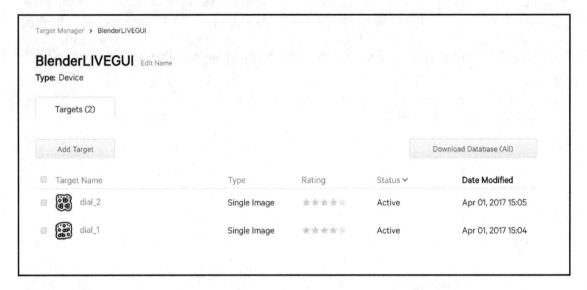

We now have our targets but have yet to add the SDK to use of them; let's do that now. Near the top of the page, click on **Downloads**. This will take you to the SDK section of the available downloads. Click on the **Download for Unity** link and accept their terms and conditions to begin downloading. Once downloaded, it is simply a matter of importing the package, either by double-clicking on the package within Windows Explorer or importing from within the Unity Editor using the menu **Assets | Import Package | Custom Packages ...** menu item. Import the whole package or remove the dependencies of iOS and Android if you want to save a bit of room. Next, we can import our targets we downloaded earlier using the same process; either double-click on the Unity package we downloaded or import it from within the Unity Editor.

We are now ready to give our users the ability to manipulate the hologram using physical objects. Of course, you will need to print out the targets we have just created and fasten them to some physical object that can easily rotate without obstructing the target too much.

Now is probably a good time to pause for a moment and revise our goal. What we are trying to achieve is providing our users with the ability to manipulate the hologram using physical objects. Ideally (and it is feasible), we would achieve this without having any artificial artefacts (that is, our targets/markers). These physical objects will be assumed to be in close proximity to the user and, in this example, we will have two objects, one controlling rotation around the y-axis and the other controlling translation along the y-axis. The user will make adjustments (rotation or translation) by rotating the physical object. With the concept now fleshed out, it's time to make it real.

First, we will need to add **ARCamera** to the scene. This is a camera Vuforia will use to capture the camera feed from the HoloLens. Enter **ARCamera** in the **Project** panel's Search bar and, once visible, drag the prefab onto the scene. Select it from the **Hierarchy** panel to make its properties visible in the **Inspector** panel and, from within the **Vuforia Behaviour** component, click on **Open Vuforia configuration** and make the following amendments:

- Copy and paste your applications licence key into the **App Licence Key** field.
- Set **Max Simultaneous Tracked Images** to **2**; this will allow both of our targets to be recognised and tracked at the same time.
- Expand **Digital Eyewear**; select **Optical See-Through** for **Type** and **HoloLens** for **See Through Config**.
- Expand **Datasets**; check **Load <Name of Vuforia Database>**. Once checked, the **Activate** option will become visible. Check this also.
- Expand Webcam and check **Disable Vuforia Play Mode**.

Return to the properties of **ARCamera** by selecting the associated GameObject from within the **Hierarchy** panel; with the properties now visible in the **Inspector** panel, make sure the **Vuforia Behaviour** component's **World Center Model** is set to CAMERA (this determines how the camera updates its position) and, finally, click and drag **Main Camera** from the **Hierarchy** panel onto the **Center Anchor Point** field below (reference camera).

Now we have our camera in place, our next task is to add the targets we're interested in tracking. Vuforia has made this easy for us by providing a prefab with most of the work done. From within the **Project** panel, type **ImageTarget** into the search box and once it's visible click and drag the prefab into the Scene. We will need a prefab for each target; I'll walk through setting up one and leave the second as an exercise for you.

Click the newly added GameObject (**ImageTarget**) from within the **Hierarchy** panel to bring up its properties in the **Inspector** panel and make the following changes to the **Image Target Behaviour** component:

- Select <**Name of Vuforia Database**> from the **Database** drop-down.
- Next select your first target (I have called mine **dial_1** and the other **dial_2**) from the **Image Target** dropdown. Once set, the **Width** and **Height** will be automatically populated.
- Uncheck **Enable Extended Tracking** if checked.
- Uncheck **Enable Smart Terrain** if checked.
- Remove the **Default Trackable Event Handler** component; we will build our own version of this script.

If the GameObject **SmartTerrain_ImageTarget** in present in your scene, delete it. This, as the name suggests, is specifically for the `SmartTerrain` functionality Vuforia offers, a feature less applicable to the HoloLens.

It's worth noting that, for our example, we have disabled Extended Tracking; Vuforia introduced this as a way to take advantage of the solid tracking HoloLens provides. Without HoloLens, Vuforia handles tracking by performing fairly expensive processing on each camera frame. Extending Tracking transfers this responsibility to the HoloLens, leaving Vuforia only concerned with the recognition and the hand-off of the targets.

We have disabled it in this example, leaving Vuforia responsible for both, recognition and tracking. We have done this for two main reasons; the first is that HoloLens tracking performs poorly with anything less that 0.8 meters (the distance our objects will likely be at); the second is that we want to detect changes in rotation and, during development, I found this wasn't detected when Extended Tracking was enabled.

If you look at your Scene, you will be presented with a white plane where your target is located. This is due to the malformed import setting on the associated textures. Although just aesthetics, it provides a useful visual aid when navigating around your scene. You can resolve this by extending `Editor/QCAR/ImageTargetTextures/<Name of Vuforia Database>/` from within your **Project** panel. For each texture, update the following properties:

- Set **Texture Type** to `Default`
- Set **Texture Shape** to `2D`

Once updated, click on the **Apply** button and, once applied, you should see the update in your scene (if not, then try selecting it from within the **Hierarchy** panel).

The next step is to add some visual feedback for the user, that is, a virtual representation of the dial. This allows us to provide feedback for the current state (degrees of change in our case) as well as keeping the user informed about the tracking state for each image. Within the `App/Models/` folder of the **Project** panel, there is a model that we will use. Once located, click and drag onto **ImageTarget** we have just created, ensuring its **Position** is set to **X:** 0, **Y:** 0, **Z:** 0 and **Rotation** is set to **X:** -90, **Y:** 0, **Z:** 0. This is a simple model made up of two parts, a circle and an arrow, with the intention of portraying the current offset using the arrow to point in the GameObjects, initial forward direction. Your project should look similar to the following; now is a good time to build and deploy to your HoloLens to test out the tracking:

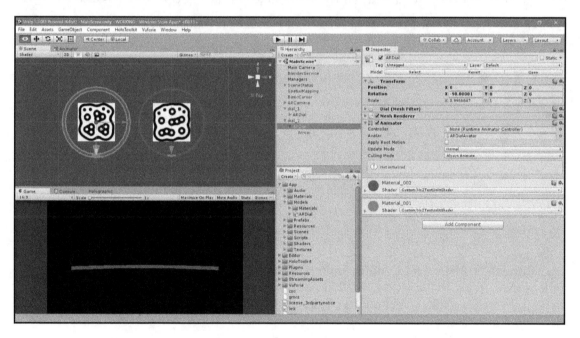

Of course, nothing happens when you rotate the targets. This is our next task. For this, we will create a script, to be attached to each target, that will be responsible for detecting and broadcasting changes in rotation. `SceneManager` will observe these events and, when detected, send the appropriate operation to the BlenderLIVE service.

Create a new script by clicking on **Create** from within the **Project** panel and selecting **C# Script**. Give the script the name `ARUIDial` and double-click to open it in Visual Studio. First, we will need to include the Vuforia namespace. Add the following at the top of the script:

```
using Vuforia;
```

Next, have the class implement the interface `ITrackableEventHandler`. This interface has a single method that Vuforia will call when the state of tracking changes; let's also add this:

```
public class ARUIDial : MonoBehaviour, ITrackableEventHandler
{
    void Start ()  {

                 }

    void Update () {

                 }

    public void OnTrackableStateChanged(TrackableBehaviour.Status
    previousStatus, TrackableBehaviour.Status newStatus)
    {

    }
}
```

As mentioned earlier, we will notify interested parties of the changes via an event which will hold reference to the `ARUIDial` that raised the event and changes, such as the change in degrees the object had been rotated. Add the following delegate and event at the top of the `ARUIDial` class:

```
public delegate void ARUIDialChanged(ARUIDial dial, float change);
public event ARUIDialChanged OnARUIDialChanged = delegate { };
```

In order to be useful, ARUIDial needs to be aware whether its currently being tracked or not. To do this, we will register for tracking events using the TrackableBehaviour component attached to the same GameObject. Because we will be frequently referencing TrackableBehaviour, we will use a variable to cache its reference. Add the following variable to the ARUIDial class:

```
private TrackableBehaviour trackableBehaviour;
```

And now, add the following code to the Start method that will register itself to receive tracking events (and obtain a reference to TrackableBehaviour):

```
void Start()
{
    trackableBehaviour = GetComponent<TrackableBehaviour>();

    if (trackableBehaviour)
    {
        trackableBehaviour.RegisterTrackableEventHandler(this);
    }
}
```

We will receive tracking events via the OnTrackableStateChanged method. In addition, we will create a property to provide us with a convenient way of knowing if we are currently tracking the target or not. Add the following property to the ARUIDial class:

```
public bool IsTracking
{
    get
{
    if (trackableBehaviour)
{
        return trackableBehaviour.CurrentStatus ==
        TrackableBehaviour.Status.DETECTED ||
        trackableBehaviour.CurrentStatus ==
        TrackableBehaviour.Status.TRACKED ||
        trackableBehaviour.CurrentStatus ==
        TrackableBehaviour.Status.EXTENDED_TRACKED;
}

    return false;
}
}
```

The term tracking refers to a concept in computer vision where you have recognized an object and are able to identify and report on its current location.

The reason we are interested in knowing when the target is being tracked or not is mainly concerned with showing or hiding the visual components of the target, in our case, the model of a dial. When `TrackableBehaviour` notifies us of a state change, we will simply show or hide any `MeshRenderer` attached accordingly. The obvious place for this code is the callback itself. Make the following changes to the `OnTrackableStateChanged` method:

```
public void OnTrackableStateChanged(TrackableBehaviour.Status
previousStatus, TrackableBehaviour.Status newStatus)
{
    if(IsStatusApproxTracking(previousStatus) ==
    IsStatusApproxTracking(newStatus))
    {
        return; // no substantial change, so ignore
    }

    if (IsStatusApproxTracking(newStatus))
    {
        OnTrackingFound();
    }
    else
    {
      OnTrackingLost();
    }
}
```

Here, we are calling a helper method that generalises the state, as we did with the `IsTracking` property . If there is no change, then we simply ignore the request; otherwise, we call `OnTrackingFound` when tracking and `OnTrackingLost` when tracking has been lost. Let's add all three methods (`IsStatusApproxTracking`, `OnTrackingFound`, and `OnTrackingLost`). Now, add the following code to the `ARUIDial` class:

```
private bool IsStatusApproxTracking(TrackableBehaviour.Status status)
{
return status == TrackableBehaviour.Status.DETECTED ||
status == TrackableBehaviour.Status.TRACKED ||
status == TrackableBehaviour.Status.EXTENDED_TRACKED;
}
```

Similar to our `IsTracking` property, whenever the target has been identified it is considered as being, as we are not too concerned with the details:

```
private void OnTrackingFound()
{
  Renderer[] rendererComponents = GetComponentsInChildren<Renderer>
  (true);
  Collider[] colliderComponents = GetComponentsInChildren<Collider>
  (true);

  foreach (Renderer component in rendererComponents)
  {
    component.enabled = true;
  }

  foreach (Collider component in colliderComponents)
  {
    component.enabled = true;
  }
}

private void OnTrackingLost()
{
  Renderer[] rendererComponents = GetComponentsInChildren<Renderer>
  (true);
  Collider[] colliderComponents = GetComponentsInChildren<Collider>
  (true);

  foreach (Renderer component in rendererComponents)
  {
    component.enabled = false;
  }

  foreach (Collider component in colliderComponents)
  {
    component.enabled = false;
  }
}
```

Here, we implement the methods responsible for enabling and disabling all the attached and nested `Colliders` and `Renderers`.

The last task for the `ARUIDial` is to monitor and broadcast rotational changes, along with updating its visual state to make it more apparent to the user what is happening. For this example, we have taken a very simplistic approach; when first detected, we will remember the GameObject's current forward and up direction.

If any significant change is detected in the up direction, we will assume the user is moving the object and ignore the change; otherwise, we will compare the current forward direction for `transform` with the previous and use this to determine any change. Let's do this now; start by adding the following variables to the `ARUIDial` class:

```
private bool initilised = false;

private Vector3 previousUp = Vector3.zero;
private Vector3 previousForward = Vector3.zero;
```

We will use `previousUp` and `previousForward` to store the previous directions of the GameObject for each update and use `initilised` to flag when we are ready to make the comparison; all of this will occur within the `Update` method as such. Make the following amendments:

```
void Update()
{
if (IsTracking)
{
if (initilised)
{
if (Vector3.Dot(previousUp, transform.up) < 0.8f)
{
initilised = false;
}
else
{
float change = Vector3.Angle(transform.forward, previousForward);
Vector3 cross = Vector3.Cross(transform.forward, previousForward);
if (cross.y < 0) change = -change;

if (Mathf.Abs(change) > 5f)
{
OnARUIDialChanged(this, change);
previousForward = transform.forward;
}
}
}
else
{
initilised = true;
previousUp = transform.up;
previousForward = transform.forward;
}
}
}
```

As mentioned earlier, we are only concerned when the object is being tracked and, if it is being tracked, we first test the current and previous up direction of the GameObject's `transform`. If we detect a large change, then we assume tracking is unstable and flag `initialised` to false to force the script to reinitialize the up direction. If considered stable (the `transform` up direction hasn't changed significantly), we compare the `transform` GameObject's current forward direction with the previous forward direction, and use the difference to determine the change in angles. If this change is greater than 5 degrees, we make the subscribed listeners aware by firing the `OnARUIDialChanged` event.

If not initialized, we simply update `previousUp` and `previousForward` for processing in subsequent updates.

Thus far, we have detected the changes but provided no feedback to the user. For this, we will leave the outer ring to rotate freely with the target (the direction the user is rotating) but keep the child GameObject (arrow) locked to the initialized forward direction. First things first, we need a reference to the GameObject we will be locking and a variable to store the rotation on initialization. At the top of the `ARUIDial` class, add the following public property:

```
public GameObject lockedGameObject;
private Quaternion lockedGameObectsRotation = Quaternion.identity;
```

We will assign the nested object to this variable once back in the Editor; for now, we will continue updating the code. Our next task is to keep track of the rotation on initialization; as you would expect, this is done when setting `initialised` to true. Jump back into the `Update` method and make the following amendment:

```
void Update()
{
if (IsTracking)
{
if (initilised)
{
// detect displacement by checking the current up with the previous
if (Vector3.Dot(previousUp, transform.up) < 0.8f)
{
initilised = false;
}
else
{
float change = Vector3.Angle(transform.forward, previousForward);
Vector3 cross = Vector3.Cross(transform.forward, previousForward);
if (cross.y < 0) change = -change;

if (Mathf.Abs(change) > 5f)
```

```
{
OnARUIDialChanged(this, change);
previousForward = transform.forward;
}
}
}
else
{
initilised = true;
previousUp = transform.up;
previousForward = transform.forward;

if (lockedGameObject)
{
lockedGameObectsRotation = lockedGameObject.transform.rotation;
}
}
}
}
}
```

Here, we are simply caching the rotation of the assigned lockedGameObject. Next, we will force this rotation by continuously resetting the lockedGameObjects rotation to this cached value. To avoid conflicts with Vuforia trying to update its rotation, we will perform its update in the LateUpdate method, essentially allowing us to have the last say. Add the following to the ARUIDial class:

```
private void LateUpdate()
{
    if (IsTracking)
     {
       if (initilised)
    {
       if (lockedGameObject)
    {
       lockedGameObject.transform.rotation = lockedGameObectsRotation;
    }
    }
    }
}
```

Here, we are simply checking whether we are currently tracking the target and are a valid state before resetting the `lockedGameObject.transform.rotation` with the cached value. This now concludes our `ARUIDial` script. Jump back into the Unity Editor and attach this script onto each of our targets by simply clicking on each from within the **Hierarchy** panel, then clicking on the **Add Component** button from the **Inspector** panel, and typing `ARUIDial`. With the script attached to both, expand each GameObject in the **Hierarchy** panel and drag the **Arrow** GameObject onto the **Locked Game Object** field. After this, your target GameObject properties should look similar to the screenshot shown:

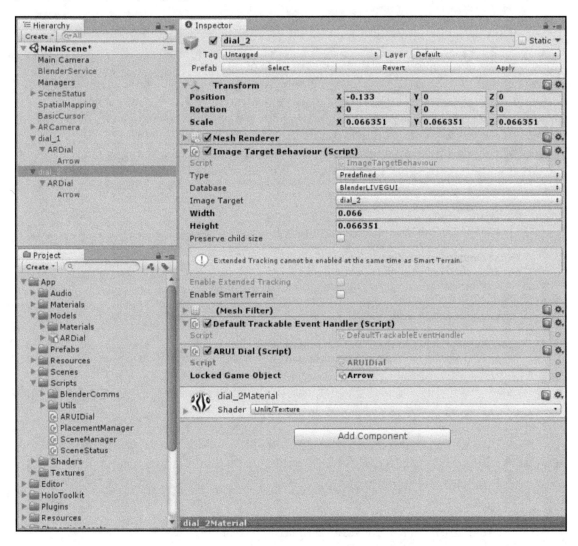

With our targets now wired up, the final task is to subscribe to the associated events and relay the adjustments to `BlenderLiveManager`. As mentioned earlier, this will be the responsibility of `SceneManager`. So, with that in mind, let's open `SceneManager` in Visual Studio and make it responsible. Start off by adding the following instance variables:

```
public ARUIDial rotationUIDial;

public ARUIDial translateXUIDial;

public float rotationPerDegree = 1.0f;

public float metersPerDegree = 0.00138f;
```

Here, we have declared two variables, one for each of our targets; `rotationUIDial` for rotating and `translateXUIDial` for translating the hologram (as discussed earlier). Next, we declare `rotationPerDegree`, which defines the transformative relationship between the physical and virtual object. Setting it to 1 means that a change of 1 degree of `rotationUIDial` will change the hologram by 1 degree. Similarly, with `metersPerDegree`; this variable defines the relationship between a change in the degree of the target with the corresponding translation. We will register for the relevant event in the `Start` method:

```
void Start () {
if (rotationUIDial)
{
  rotationUIDial.OnARUIDialChanged +=
  RotationUIDial_OnARUIDialChanged;
}

if (translateXUIDial)
{
  translateXUIDial.OnARUIDialChanged +=
  TranslateXUIDial_OnARUIDialChanged;
}

...
}
```

And finally, we will write the code to handle each of the events, starting with the event associated with `rotationUIDial`. Add the following method to the `SceneManager` class:

```
private void RotationUIDial_OnARUIDialChanged(ARUIDial dial, float
change)
{
   float rotation = rotationPerDegree * change;
```

```
    var bgoNames =
    BlenderServiceManager.Instance.GetAllBlenderGameObjectNames();

    if (bgoNames.Count == 0)
    {
        return;
    }

    var bgo =
     BlenderServiceManager.Instance.GetBlenderGameObjectWithName(bgoNames[0
    ]);
       BlenderServiceManager.Instance.SendOperation(bgo,
       BlenderServiceManager.ObjectOperations.Rotate, new Vector3(0, 0,
       rotation));
    }
```

When the associated target is rotated, it broadcasts the event passing along change in degrees. We use this with the associated weight to determine the change to be applied to the imported Blender object. After this, we verify that we have at least one Blender object loaded (in this example, we are assuming only a single model will be loaded). After obtaining a reference to the first Blender object, we will send the request by calling SendOperation of BlenderServiceManager. Once the BlenderLIVE service receives the operation, it will execute the operation locally and broadcast the changes to all the connected clients. We do exactly the same for the translation dial but using a different weight and operation. Add the following method to your SceneManager class:

```
private void TranslateXUIDial_OnARUIDialChanged(ARUIDial dial, float
change)
{
    float displacement = metersPerDegree * change;

    var bgoNames =
    BlenderServiceManager.Instance.GetAllBlenderGameObjectNames();

    if (bgoNames.Count == 0)
    {
        return;
    }

var bgo =
BlenderServiceManager.Instance.GetBlenderGameObjectWithName(bgoNames[0 ]);
BlenderServiceManager.Instance.SendOperation(bgo,
BlenderServiceManager.ObjectOperations.Translate, new Vector3(0,
displacement, 0));
}
```

With this in place, we now have realized the functionality to manipulate the hologram using physical objects. Build, deploy, and test to see it in action before returning, where we talk about how we can create a shared virtual environment between multiple users:

Shared space

The HoloLens uses a concept known as spatial anchors to map out the environment. These anchors are reference points used by the device to build a coordinate system and used to position holograms. Using multiple anchors allows the device to track large areas. These anchors can also be persisted and shared; sharing an anchor, along with its supporting understanding of the environment, is what we will cover in this section.

Just to recap, our goal is to allow multiple users to review a design outside the screen. To enable this, we need to have a single user place the hologram and share this location with other users that join, such that all users are viewing the hologram from the same place.

We are already placing the hologram, so our first task will be extending this such that when the hologram is initially placed, we attach a `WorldAnchor` to it and, once attached, export and transfer the data to the BlenderLIVE service where it can be shared with subsequent users that join. After finishing this, we will look at how we can import an existing `WorldAnchor` to automatically place the hologram.

Jump into the `SceneManager`, **navigate to the** `PlacementManager_OnObjectPlaced`
method, and make the following amendments:

```
    private void PlacementManager_OnObjectPlaced(BaseBlenderGameObject bgo)
    {
    SceneStatus.Instance.SetText("");

    BaseBlenderGameObject.Anchored onAnchoredHandler = null;
    onAnchoredHandler = (BaseBlenderGameObject caller, byte[] data) =>
    {
    caller.OnAnchored -= onAnchoredHandler;
    BlenderServiceManager.Instance.SendBlenderGameObjectWorldAnchor(caller,
    data);
    };

    bgo.OnAnchored += onAnchoredHandler;
    bgo.AnchorAtPosition(bgo.transform.position);
    }
```

Creating an anchor in Unity is simply a matter of attaching a `WorldAnchor` component to
the GameObject you want to anchor, which we will do via the `AnchorAtPosition` method
next. Our main intention in creating an anchor is to create a shared space. For this, we need
to export and share it. This is an asynchronous task, which we will implement shortly in the
`AnchoredBlenderGameObject` class. Once exported, we will fire the `OnAnchored` event,
which we are registering here, passing the serialized data. Once received, we pass it to the
BlenderLIVE service so that it can be shared with the other users that join our session.

Open the script `AnchoredBlenderGameObject`, navigate to the comment `// TODO
Implement exporting the WorldAnchor` within the `AnchorAtPosition` method, and
make the following amendments:

```
    public override bool AnchorAtPosition(Vector3 position)
    {
    if (exporting || importing)
    {
    return false;
    }

    if (worldAnchor)
    {
    Destroy(worldAnchor);
    }

    gameObject.transform.position = position;

    #if WINDOWS_UWP
```

```
worldAnchor = gameObject.AddComponent<WorldAnchor>();

if (worldAnchor.isLocated)
{
StartExportingWorldAnchor();
}
else
{
worldAnchor.OnTrackingChanged += WorldAnchor_OnTrackingChanged;
}

exporting = true;

#endif

return true;
}
```

As mentioned earlier, in Unity creating an anchor is as simple as adding the
WorldAnchor component to the GameObject. After attaching the WorldAnchor, we
determine whether the anchor is located or not; in other words, is HoloLens currently
tracking the object or not. If so, we call StartExportingWorldAnchor which, as the name
suggests, takes care of exporting the WorldAnchor. If not located, we register for the
OnTrackingChanged events and patiently wait for the anchor to be tracked before calling
StartExportingWorldAnchor to kick off the process. Let's now implement
the WorldAnchor_OnTrackingChanged method. Add the following method to the
AnchoredBlenderGameObject class:

```
private void WorldAnchor_OnTrackingChanged(WorldAnchor self, bool located)
  {
    if (located)
    {
      worldAnchor.OnTrackingChanged -= WorldAnchor_OnTrackingChanged;
      StartExportingWorldAnchor();
    }
  }
```

An `WorldAnchor` is a serialized description of a specific point in space, including supporting environment data; anchors can be fairly large packets of data. To export, we call the `ExportAsync` method of the `WorldAnchorTransferBatch`. As the name suggests, the exporting takes places asynchronously using the anchor (or anchors) batched inside the `WorldAnchorTransferBatch` class along with two delegates that are notified as data becomes available and when the task is finished. The serialized batched data is passed as chunks to the `SerializationDataAvailableDelegate` handler; here, we will cache each chunk and, once finished, pass the full array of data to the `OnAnchored` event. For caching, we will simply use a generic list using bytes arrays as the generic. Let's add this variable now to the `AnchoredBlenderGameObject` class:

```
private List<byte> worldAnchorBuffer = new List<byte>();
```

And now, let's implement the method responsible for preparing and starting the export. Add the following to the `StartExportingWorldAnchor` method:

```
private void StartExportingWorldAnchor()
{
    if (worldAnchor == null)
    {
        worldAnchor = gameObject.GetComponent<WorldAnchor>();
    }

    if (worldAnchor == null)
    {
        return;
    }

    worldAnchorBuffer.Clear();

    WorldAnchorTransferBatch transferBatch = new
    WorldAnchorTransferBatch();
    transferBatch.AddWorldAnchor(gameObject.name, worldAnchor);
    WorldAnchorTransferBatch.ExportAsync(transferBatch,
    OnExportDataAvailable, OnExportComplete);
}
```

Here, we are clearing our buffer of any previous exports and creating
`WorldAnchorTransferBatch`, a structure simply used to bundle `WorldAnchors` together
for exporting. We add the `WorldAnchor` we want to export using the `AddWorldAnchor`
method and being exported by calling `ExportAsync`, passing in our batch and handlers,
currently missing at present. Let's fix that now by adding both to our
`AnchoredBlenderGameObject` class, starting with the `OnExportDataAvailable`
method:

```
private void OnExportDataAvailable(byte[] data)
{
    worldAnchorBuffer.AddRange(data);
}
```

As mentioned earlier, `WorldAnchorTransferBatch` will pass through chunks of the
serialised anchor. Here, we are simply caching them in our `worldAnchorBuffer` list.

The `SerializationCompleteDelegate` handler is called once exporting is complete.
Complete here means having finished the task rather than having successfully exported the
`WorldAnchor`. It encodes the result of the export using the
enumeration `SerializationCompletionReason`. If its value is equal to `Succeeded`, then
we know we have successfully exported the anchor; otherwise, we need to handle the
exception (omitted in this example as with most exception handling):

```
private void OnExportComplete(SerializationCompletionReason
completionReason)
{
    exporting = false;

    if (completionReason == SerializationCompletionReason.Succeeded)
    {
        RaiseOnAnchored(worldAnchorBuffer.ToArray());
    }
    else
    {
        // TODO: handle expectational case
    }
}
```

Once we have successfully completed exporting, we call `RaiseOnAnchored`, passing
along the cached data. This, in turn, will fire `OnAnchored`, which `SceneManager` is eagerly
waiting for.

This now completes the setting and exporting of the `WorldAnchor`, which other HoloLens users can import so everyone is working in a shared space. The following figure summarizes the main steps of the process:

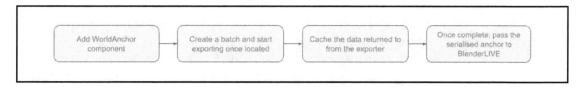

Our next task is to handle the other side, that is, importing. This occurs when we receive a Blender object with an existing `WorldAnchor`. Similar to exporting, the entry point of this process begins in `SceneManager` with the bulk of the logic encapsulated in the `AnchoredBlenderGameObject` class.

 The process of storing and transferring this data is abstracted away in the `BlenderServiceManager` and associated classes; I encourage you to explore these to get a better understanding of how this is handled.

There are two cases when we need to check whether a received Blender object has an associated anchor. The first time a Blender object is retrieved (created) as well as every time the object is updated. Starting with the creation use case, back in `SceneManager`, locate the method `BlenderServiceManager_OnBlenderGameObjectCreated` and make the following amendments:

```
private void BSM_OnBlenderGameObjectCreated(BaseBlenderGameObject bgo)
{
if (bgo.IsAnchored)
{
SceneStatus.Instance.SetText("");

if (bgo.GetComponent<WorldAnchor>())
{
if (PlacementManager.Instance.RemoveObjectForPlacement(bgo.name))
{
SceneStatus.Instance.SetText("Blender object placed", MEDIUM);
}
}
}
else
{
SceneStatus.Instance.SetText("Blender object ready for placementnAir-tap
on a suitable surface to place", MEDIUM);
PlacementManager.Instance.AddObjectForPlacement(bgo);
```

```
        }
    }
```

Previously, we were only concerned with passing the created object to `PlacementManager`. `PlacementManager` would then take responsibility for placing the object. This time, we first check whether the object has a `WorldAnchor`. If not, then we follow the same path as we had previously, but if the object does have a `WorldAnchor`, we first ensure that `PlacementManager` doesn't have the object in its queue by calling the method `RemoveObjectForPlacement` and then simply notifying the user that the object has been placed. Placement occurs when the associated `WorldAnchor` is imported and attached to the GameObject. This occurs when the serialised data transmitted from the BlenderLIVE service is bounded to the GameObject, which in turn calls the `SetAnchor` method of the `AnchoredBlenderGameObject` class, passing it the serialized `WorldAnchor`. It is here we will start with our next chunk of amendments. Make the following amendments to the `SetAnchor` method:

```
public override bool SetAnchor(byte[] data)
{
if (exporting || importing)
{
return false;
}

#if WINDOWS_UWP

importing = true;

worldAnchorBuffer.Clear();
worldAnchorBuffer.AddRange(data);

importAttempts = WorldAnchorImportAttempts;
WorldAnchorTransferBatch.ImportAsync(data, OnImportComplete);

#endif

return true;
}
```

We first check to see whether we are currently either importing or exporting. If so, we return; otherwise, we continue the importing process. Within this class, we have defined a constant called `WorldAnchorImportAttempts` and the `importAttempts` variable. The former defines the import attempts we will try before, essentially, giving up. The latter, `importAttempts`, is the current number of attempts remaining for the current import. The importing itself is abstracted away in the `ImportAsync` method of the `WorldAnchorTransferBatch` class. We simply need to pass it the serialized data and a callback delegate, which we will implement now:

```
private void OnImportComplete(SerializationCompletionReason
completionReason, WorldAnchorTransferBatch deserializedTransferBatch)
{
importing = false;

if (completionReason != SerializationCompletionReason.Succeeded)
{
if (importAttempts > 0)
{
importing = true;
importAttempts--;
WorldAnchorTransferBatch.ImportAsync(worldAnchorBuffer.ToArray(),
OnImportComplete);
}
return;
}

worldAnchor = deserializedTransferBatch.LockObject(gameObject.name,
gameObject);
}
```

First, we check to see whether the import was successful or not. If not, then we attempt it again until we have reached our threshold. Otherwise, we call the `LockObject` method of the `WorldAnchorTransferBatch` instance. We pass this method the name of the associated `WorldAnchor` and the GameObject we want be to anchored (or locked if we use their naming convention).

We have one final amendment to make before we have extended our application across multiple devices; if you remember, I mentioned that `WorldAnchor` could be attached when `BlenderGameObject` was created or updated. Here, we have taking care of cases when it's created and, now, we will handle the case for when `BlenderGameObject` is updated, which occurs if two users are connected to the BlenderLIVE service when the object hasn't yet been placed.

As soon as one user places the object, the update method will be called on the other, passing along the newly created WorldAnchor. Jump back in the SceneManager class and make the following amendments to the BlenderServiceManager_OnBlenderGameObjectUpdated method:

```
private void BSM_OnBlenderGameObjectUpdated(BaseBlenderGameObject bgo)
{
SceneStatus.Instance.SetText("");
if (bgo.IsAnchored)
{
if (PlacementManager.Instance.RemoveObjectForPlacement(bgo.name))
{
SceneStatus.Instance.SetText("Blender object placed", MEDIUM);
}
}
}
```

As we did with the BSM_OnBlenderGameObjectCreated method, we test whether an anchor exists and, if so, ensure that it is removed from the PlacementManager queue; and we finally, notify the user that the object has been placed.

With that block of code now finished, we have now extended Blender beyond the flat screen and into the real world, giving designers a better, more natural, forum to communicate and collaborate on. Grab a friend nearby with the device and build and deploy to test.

In the next section, we take advantage of spatial sound to make discovering the holograms easier.

Assisted discovery with spatial sound

So far, we have allowed two or more users to share a space by sharing a single WorldAnchor; this allows everyone to see the hologram in the same place, but there is a slight issue we need to address. The placement of a hologram may not be obvious to the user who hasn't placed it. Currently, we just tell that user the hologram has been placed; of course, given this is a collaborative tool, we can assume the user who placed the hologram will explicitly or implicitly indicate where the hologram is placed, but assumptions are a sure path to a poor user experience.
What are our options? We can do either of the following:

- Show a visual element to indicate the direction of the hologram if the hologram is out of view
- Make a sound in the location of the hologram

In this section, as implied by the heading, we will tackle this challenge using spatial sound. But it's worth taking some time to briefly introduce spatial sound and acknowledge its importance for creating an immersive experiences.

Spatial sound provides the ability to simulate sound in a 3D environment, adjusting how the sound is perceived by the source direction and distance, complementing the holograms and making them seem more believable. Some common use cases for spatial sound include the following:

- **Increasing discoverability of holograms**: just as we propose to do here, use spatial clues to help the user locate a hologram.
- **Audio haptics**: an example here is using it as feedback to acknowledge a user action, such as selecting a hologram.
- **Increasing immersion**: it seems too obvious to highlight but we experience our world through all of our senses. Including more of these is going to more effectively immerse the user into the experience.

By now, hopefully, you're convinced how important spatial sound is in creating a immersive MR experience and here we are just scratching the surface. With that being said, let's put the theory into practice.

Our approach will consist of a proxy, placing it in the location of the placed hologram before playing the sound. This will help the user locate a hologram they havn't placed.

Let's start by creating a proxy. This proxy will have an `AudioSource` attached when an object is placed (either by the user or by the shared `WorldAnchor`). We will relocate this proxy to where the object was placed the hologram and play a sound effect to indicate it has been placed. In the Unity Editor, create a new empty GameObject using the menu item **GameObject | Create Empty** and give it the self describing name **PlacementSoundEffect**. With the new empty GameObject selected, navigate over to the **Inspector** panel and click on the Add **Component** button. Type and select `AudioSource` and make the following changes:

- Check **Spatialize**
- Uncheck **Play on Awake**
- Drag the dial of Spatial Blend all the way to the right (3D)

You might find that your `AudioSource` is missing **Spatialize**. This is disabled by default; follow these steps to enable it:

- Open the Audio project settings via the menu **Edit** | **Project Settings** | **Audio**
- Set **Spatializer Plugin** to **MS HRTF Spatializer**

With our `AudioSource` now configured, let's assign the `AudioClip`. Within the **Project** panel, enter **Placed** into the search field. Once visible, click and drag onto the `AudioSource--AudioClip` field of `PlacementSoundEffect`. We are now ready to make some sound; open `SceneManager` in Visual Studio and make the following changes.

First we will add a public variable gain a reference to our audio proxy:

```
public GameObject placementSoundEffect;
```

Next, we will create a helper method that will be responsible for playing the sound:

```
private void PlayPlacementSoundEffect(Vector3 position)
{
if (placementSoundEffect)
{
placementSoundEffect.transform.position = position;
placementSoundEffect.GetComponent<AudioSource>().Play();
}
}
```

Our method takes in the position of where we want the sound to be located and calls `Play` on the attached `AudioSource` of our audio proxy.

There are three places where the object, potentially, gets placed; for each path we will make a call to `PlayPlacementSoundEffect`. The first method we will amend is the delegate handling object placement called by the `PlacementManager`. Locate `PlacementManager_OnObjectPlaced` and make the following amendments:

```
private void PlacementManager_OnObjectPlaced(BaseBlenderGameObject bgo)
{
...

...
PlayPlacementSoundEffect(bgo.transform.position);
}
```

Our last two methods are when a Blender object is created or updated as each of these could contain a `WorldAnchor`. Make the following amendments to both:

```
private void BSM_OnBlenderGameObjectCreated(BaseBlenderGameObject bgo)
{
if (bgo.IsAnchored)
{
SceneStatus.Instance.SetText("");

if (bgo.GetComponent<WorldAnchor>())
{
if (PlacementManager.Instance.RemoveObjectForPlacement(bgo.name))
{
SceneStatus.Instance.SetText("Blender object placed", MEDIUM);
PlayPlacementSoundEffect(bgo.transform.position);
}
}
}
else
{
SceneStatus.Instance.SetText("Blender object ready for placementnAir-
tap on a suitable surface to place", MEDIUM);
PlacementManager.Instance.AddObjectForPlacement(bgo);
}
}

private void BSM_OnBlenderGameObjectUpdated(BaseBlenderGameObject bgo)
{
SceneStatus.Instance.SetText("");

if (bgo.IsAnchored)
{
if (PlacementManager.Instance.RemoveObjectForPlacement(bgo.name))
{
SceneStatus.Instance.SetText("Blender object placed", MEDIUM);
PlayPlacementSoundEffect(bgo.transform.position);
}
}
}
```

It's that simple. Congratulations! This now concludes this example. As usual, build and deploy to the device to marvel at your hard work and, once finished, we will wrap up with a quick summary of what you've covered in this chapter.

Summary

In this chapter, our goal was to explore the concept of creating a more natural collaborative experience for designers to review and critique their work. We did this by pulling objects from the 3D package Blender and bringing them into the real world. We then integrated Vuforia to allow the user to manipulate the hologram using physical objects, by rotating a couple of dials, where one rotates the object and the other moves it. We next looked at how we could create a shared space by sharing a `WorldAnchor` with all users. This required exporting it from a single user and having the other users import it. This allowed each device to have a shared understanding of the environment such that we could place a hologram in a place that was in the exact same location across all devices. We wrapped up this chapter by using spatial sound to help our users locate holograms when placed.

We have only just scratched the surface of each of these topics but I hope it is enough to excite you to explore what else is possible. In the next chapter, we will wrap up our chapters on Unity by building a simple but fun multiplayer game inspired by the, now classic, game Cannon Fodder (concept at least).

8
Developing a Multiplayer Game Using Unity

In the earlier chapters, we covered ways of understanding the environment, how to interact with holograms, how to map and navigate around the real world, and how to share spatial information across devices. In this chapter, we will tie these concepts together with a fun multiplayer game, with plenty of gaps for you to improve and build on top of. Unlike the previous chapters where we deep dived into specific topics, this chapter differs with the intention of quickly walking through building a simple multiplayer game, illustrating the general steps and covering any relevant considerations.

In this chapter we will cover:

- Introduction to some core concepts and components used to make multiplayer games in Unity
- Specifically looking and how we can discover peers on the same network and communicate with them
- Finally looking at how we can share the same coordinate space allowing for a shared user experience

By the end of this chapter, you will have a good sense of what is required when making a multiplayer game. With that being said, let's jump into the world of networked gaming.

Multiplayer networking

In this section, we will briefly introduce networking, specifically networking in the context of multiplayer games and Unity. It's worth emphasizing the word *briefly* as it's not the intention of this section (and chapter) to teach you everything you need to know to create a multiplayer game, but to build up a vocabulary of meanings and concepts so that this chapter makes sense. For more comprehensive details, refer to the official Unity documentation at `https://docs.unity3d.com/Manual/UNet.html` or dedicated books, such as *Unity Multiplayer Games* by Alan R. Stagner from Packt Publishing, which can be referred to at `https://www.packtpub.com/game-development/unity-multiplayer-games`.

Let's start by introducing the main components and how their architecture differs from single-player games. At its core, a multiplayer networked game (from here on referred to as multiplayer game for brevity) consists of components responsible for the following:

- Discovery
- Communication
- Coordination

Discovery is concerned with finding existing games; this can be done through a central repository that tracks the current games or something more dynamic, where each game session advertises itself and is discovered through intentional inspection.

Communication refers to the task of sending and receiving data to and from peers; how it achieves this depends on the topology and protocol. The most typical topology, and one you are no doubt familiar with, is a server and client. This topology is what we use when browsing the internet--the server is responsible for serving multiple clients, where the clients only know of each other through the server (if at all). The alternative to this is peer-to-peer, where each participant communicates with each other (and many variants in-between). Protocol here refers to the way in which data, raw bytes, is transmitted to and from participants. The two main protocols are **User Datagram Protocol** (**UDP**) and **Transmission Control Protocol** (**TCP**), where UDP is more analogous to firing an email/text/message to someone (connection-less) while TCP is analogous to calling someone on the phone (connection-oriented).

Lastly, coordination is concerned with managing and synchronizing the shared state of the game; this includes networked objects (such as player objects) and game states. It simplifies things if we have a single participant managing a specific object, which avoids having conflicts in their states, that is, allowing the associated virtual player to be controlled only by the participant who created it rather than everyone in the game. Once updated, the entity needs to be synchronized with its counterparts across the network; this is normally done by serializing its state and sending it to any connected peers to have their instances updated.

There are a lot of moving parts and luckily Unity has encapsulated a lot of this in an API called the **High Level API** (**HLAPI**). It literately consists of a set of components that handle the aforementioned tasks. We will return to these concepts and components as we walk through building up the example for this chapter, but before that, let's introduce what we plan to build.

 Along with HLAPI, Unity also exposes Transport Layer API for more control. HLAPI is sufficient for our example and will be the API we will focus on. It is also worth noting that Unity has made this API open source, and it is available at `https://bitbucket.org/Unity-Technologies/networking`.

The project

As a kid, I spent countless hours playing *Cannon Fodder*, a war-themed action game developed by Sensible Software. As a player, you controlled a squadron of soldiers to complete a mission. The controls were simple--click on an area you would like your squad to go to. If any enemies were spotted, your squad would happily open fire. As you progressed through the levels, you had the opportunity to have your squad jump into military vehicles.

The following is the screenshot of the game Cannon Fodder:

Screenshot from the 1993 computer game Cannon Fodder developed by Sensible Software, Source: Screenshot.

It is this game that inspired this example, but the environment through which you navigate your squad is replaced with the real world instead of a virtual one! Obviously, this example does not do this game justice, but I think you will agree that, as a concept, this could be an entertaining game. In addition to navigating around the real world, I swapped out the concept of war with the more friendly game of paintball, that is, instead of lead bullets, there are balls of paint, and instead of death sometimes a painful bruise. The following image is a rendering of one of the squad players we will be navigating through the environment:

Despite the differences, we do borrow the core mechanic of point and click. Here, the user points via their gaze and clicks using the tap gesture. In the next section, we will review the starter project and then start building up the project.

Project setup

In this section, we will walk through the details of the starter project and accompanying components before extending it across the network. If you haven't already, clone or download the repository for this book from `https://github.com/PacktPublishing/Microsoft-HoloLens-By-Example`. The project we will be working on is located in the `Starter` folder within the `Chapter8/Starter` directory. Once cloned, launch Unity and load the Unity project. Once loaded, you should see the familiar setup and components we have been using throughout this book:

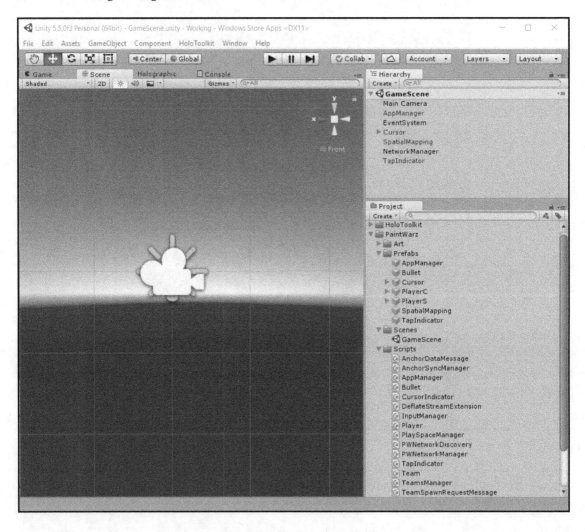

To help you get acquainted with the project so that you can confidently navigate around, we will spend the rest of this section walking through the scene components and project files:

- `MainCamera`: Something you will be familiar with now, this is our Holographic camera.
- `AppManager`: This GameObject comprises of the component related to managing the game, including some related to user interaction. When you inspect it, you will see the `AppManager`, `TeamManager`, `GazeManager`, `AnchorSyncManager`, and `InputManager` (script) components. `AppManager` is responsible for orchestrating the game state. `TeamManager` is responsible for keeping track of the teams (where each user will have their own team) and associated players (here, referring to the virtual characters, aka the members of the squad). `GazeManager`, which you are no doubt familiar with now, is responsible for tracking the user's gaze and is used by the `Cursor` to update its position and orientation. The `AnchorSyncManager` is something we will be working with quite a bit and is responsible for exporting and importing the class `WorldAnchor`. We will cover this further as we move ahead. Finally, we have `InputManager`, a simple script to listen out for the tap gesture, and notify its observers when detected.
- `Cursor`: The visual representation of the user's gaze--something we've used extensively throughout this book.
- `SpatialMapping`: Something you will also be familiar with, with the addition of the `PlaySpaceManager` script that is responsible for coordinating the construction of planes from the observed surfaces and notifying interesting parties (`AppManager`) when a sufficient horizontal surface area has been found.
- `NetworkManager`: This GameObject comprises of the components responsible for networking discovery and communication. We will return to this shortly to describe how it works in detail.
- `TapIndicator`: This is a simple visual element to show that the user has tapped when moving their squad around the floor.

Also noteworthy are the following prefabs:

- `Bullet`: This is the prefab used as a bullet. It's simple a cube with a `Rigidbody` and `Collider` attached, along with a script to manage its own state, and it notifies a `Player` GameObject if a collision occurs.
- `PlayerS` and `PlayerC`: These are the squad members that the user controls. `PlayerS` will be used by the user who assumes the role of the server, and `PlayerC` will be the client. Here, there's a subtle difference in texture, but it can easily be swapped out with different models.

Unlike the earlier chapters where we built up some classes from scratch, because we won't be introducing any new concepts related to HoloLens here, we will instead be making amendments to the existing scripts (places annotated a `// TODO` comment).

We will start at the bottom, implementing the networking component starting with discovery, connectivity, and then coordination.

Discovery

In this section, we will discuss the details of discovery while making the necessary amendments to our project. Discovery can be done statically or dynamically; statically here means that the server (and roles) are predetermined and set. In this case, when predefined, there is no need for discovery as it has already been set prior to runtime. In our case, we want the network and roles to be dynamic; we have the advantage of knowing that the peers are likely to be playing in close proximity of each other and most likely using the same network. Due to this, we can take advantage of Unity's `NetworkDiscovery` component. This component has two core functions, as follows:

- Broadcasting a message onto the network
- Listening out for a broadcast message

The following is a screenshot of the `NetworkDiscovery` component shown in the Unity Editor:

In this project, we have left the defaults; they allow you to configure how it broadcasts and listens and what data is encapsulated in the broadcast packet.

For this project, we have extended the `NetworkDiscovery` class, called `PWNetworkDiscovery`, found in the `/PaintWarz/Scripts/` folder within the **Project** panel. Double-click to open it up in Visual Studio.

The `PWNetworkManager` class (which we will get to next) starts discovery via the `StartScanning` method of the `PWNetworkDiscovery` class. When this is called, we begin listening to the network by calling the `StartAsClient` method of the base class and schedule a method invocation in a predefined time, as shown in the following code snippet:

```
public void StartScanning()
{
    if (!Initialize())
    {
```

```
        Debug.LogWarning("Failed to initialize NetworkDiscovery");
    }
    StartAsClient();
    Invoke("CheckConnection", connectionCheckTime);
}
```

What we are trying to do here is discover an existing game, but if no game is found, we want to set up one, that is, if no game is discovered within the time associated to the `connectionCheckTime` variable, we will start broadcasting.

If a packet is discovered, that is, if another device is broadcasting the packet on the network, the `NetworkDiscovery` class passes these details to the `OnReceivedBroadcast(string fromAddress, string data)` virtual method:

```
public override void OnReceivedBroadcast(string fromAddress, string
data)
{
    if(NetworkAddress != null)
    {
        return;
    }
    Debug.LogFormat("OnReceivedBroadcast; fromAddress: {0}, data:
    {1}", fromAddress, data);
    // TODO
}
```

When we receive a packet, we first want to see whether we have already designated a server (`NetworkAddress != null`) and ignore if this is the case. Otherwise, we want to update the address and try to connect to it.

Replace the comment `// TODO` with the following line:

```
NetworkAddress = fromAddress;
```

This calls the property that currently just updates the `_networkAddress` local variable, as illustrated:

```
public string NetworkAddress
{
    get
    {
        return _networkAddress;
    }
    set
    {
        _networkAddress = value;
        // TODO
```

```
        }
    }
```

When we have an address, we want to try and connect to it by updating
the class NetworkManager and ask it to connect as a client. Update the // TODO comment
with the following code:

```
        networkManager.networkAddress = _networkAddress;
        networkManager.StartClient();
```

This completes discovery for when there is a service running, but what happens when
nothing is detected? For this, we return to our delayed method call we saw within the
StartScanning method. Find the CheckConnection method within
the PWNetworkDiscovery class:

```
    void CheckConnection()
    {
        // TODO
    }
```

As mentioned in the preceding code; if nothing has been discovered by the time this
method is called, we want to start broadcasting the service and ask the
class NetworkManager to start as a server. Replace the preceding // TODO comment with
the following snippet:

```
        if (DiscoveredPeer)
        {
            return;
        }
        StopBroadcast();

        NetworkTransport.Shutdown();
        NetworkTransport.Init();

        StartAsServer();

        networkManager.StartServer();
```

We first check whether we have discovered a service and, if so, exit. Otherwise, we start
broadcasting via the StartAsServer method as well as asking the class NetworkManager
to start as a server via the StartServer method.

This concludes discovery; next, we will look at communication by inspecting the
class NetworkManager.

Communication

Let's now jump back into the Unity Editor, select NetworkManager via the **Hierarchy** panel, and inspect the properties shown in the **Inspector** panel:

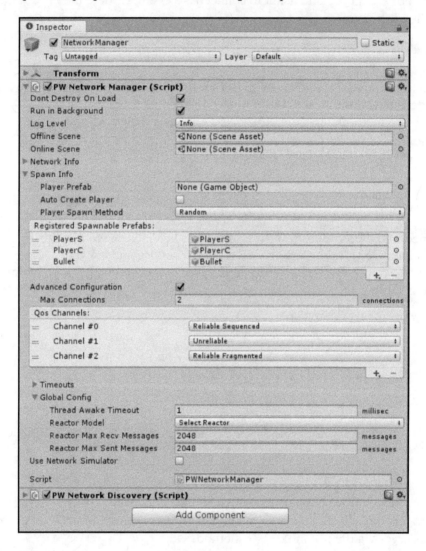

The sheer amount of properties exposed illustrates the complexities encapsulated within this class; we will now step through the relevant properties so that we have a better understanding of the NetworkManager, starting with the **Spawn Info** section:

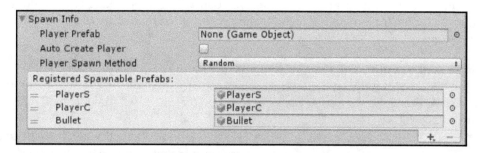

The NetworkManager is a Swiss army knife for multiplayer games; it surfaces a lot in order to hide the complexities of networking from the game developer, leaving them to concentrate on what matters, making a great game, and leaving networking to Unity. With Unity's HLAPI, it's possible to integrate networking into your game with little or no code. One of the features it offers is the ability to spawn (create) an associated player. If this is something you want, you can assign a prefab to the **Player Prefab** property (shown in the preceding screenshot) along with tweaking how it's spawned. For any other GameObject you want to spawn across the network during game play, you must register it with the NetworkManager via the **RegisteredSpawnablePrefabs**. Here, we have assigned the three prefabs we are using for this game: the two players and a bullet.

Next, we turn our attention to channels; channels in this context refers to the **Quality of Service** (**QoS**) policy applied to a message (message here meaning the structure sent between peers of a networked game); the message is still transmitted through the same connection/path as other packets, but with the addition (or removal) of checks depending on the specified channel and associated policy. From within the **Inspector** panel, and in the following screenshot, you can see the channels used in this project:

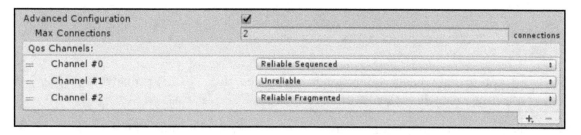

Channel #0 and **Channel #1** are the default channels assigned to the `NetworkManager`; **Channel #3** is an additional channel that has been added and will later be used to transmit the `WorldAnchor`. We have assigned the QoS policy **Reliable Fragmented,** "reliable" meaning certain mechanisms are put in place to ensure that the packet is delivered, and "fragmented" allows for a message to be broken down into small chunks and allows for larger messages.

The `PWNetworkManager` class inherits the `NetworkManager`, but offers little additions; the main motivation for extending the `NetworkManager` is due to its vast virtual methods that allow you to monitor the state of network activity better. Here, we make use of the `OnServerConnect(NetworkConnection conn)` and `OnClientConnect(NetworkConnection conn)` hooks (called when a client connects to a server and vice versa) to signal to other components that a connection has been established, using the `OnNetworkReady` and `OnConnectionOpen` events; the following is the code snippet for the method `OnClientConnect(NetworkConnection conn)`:

```
public override void OnClientConnect(NetworkConnection conn)
{
    IsServer = false;
    base.OnClientConnect(conn);
    IsReady = true;
    OnConnectionOpen(this, conn);
}
```

The following is the code snippet for the method `OnServerConnect(NetworkConnection conn)`:

```
public override void OnStartServer()
{
    IsReady = true;
}

public override void OnServerConnect(NetworkConnection conn)
{
    base.OnServerConnect(conn);
    OnConnectionOpen(this, conn);
}
```

It's worth noting that when a connection is established, the `IsReady` property is set to `true`; this is something we will make use of within the `AppManager` class.

With discovery and communication taken care of, we will now shift our focus to a discussion on how we can get the instances to use a shared space, which is the topic of our next section.

Shared space

As highlighted throughout this book, a big difference between MR and VR is the notion of space. VR being virtual, means that we easily standardize the origin of the world (normally 0,0,0); MR differs in that it uses its understanding of the physical environment to reason about where holograms should be positioned and how they are orientated, so in order for two (or more) devices to collaborate effectively, we need some way of agreeing on what the origin of the world is.

In Chapter 3, *Assistant Item Finder Using DirectX*, and Chapter 7, *Collaboration with HoloLens Using Unity*, we learned what world anchors are; we also learned how to serialize and share them. This is exactly what we need, and will use, thus allowing both devices to have a shared coordinate system of the play space--but what is our play space? In this project, we are forcing the user to place their squad on a surface close to the floor (lowest and largest plane created), so we will use this as our base and attach a WorldAnchor to the associated GameObject.

> One way of thinking about shared coordinates is if you think of you and a friend following the same map but interpreting north (forward) differently. Forward to you is different from what it is to your friend.

We need one of the devices to share their coordinate space and the other device to make use of it. In this example, we are designating the server for this task; however, as a more diverse set of devices are released, you may want to give preference to the device that can deliver better results (better here can be anything from more power efficient, higher resolution, faster, or some other heuristic) based on the logic of your game; maybe search for the smallest anchor, that is, one that takes the least amount of memory.

Before jumping into code, let's describe the steps we will follow to share and borrow the WorldAnchor; as the server and client follow different paths, we'll describe each of the steps separately, starting with the server:

- **Server process**: Once finished with scanning the environment, we attach a WorldAnchor onto the Floor and proceed to export it. While it's exporting, we cache the results passed to the OnExportDataAvailable, which will be used to share with the client. We also register a handler for a RequestAnchor message, and once received, will be sending the raw bytes of the WorldAnchor to the client.

- **Client process**: Once finished with scanning the environment and connected with the server, we request the server to send the anchor across via a `RequestAnchor` network message. Once we have received all the data, we assemble the array and begin to import the anchor, attaching it to the floor once successfully imported.

Let's begin; locate the `AnchorSyncManager` within the `PaintWarz/Scripts/` folder from within the **Project** panel and double-click on the script to open it with Visual Studio.

Locate the `PlaySpaceManager_OnPlaySpaceFinished(GameObject floorPlane)` delegate and replace the `// TODO` comment with the following snippet:

```
    private void PlaySpaceManager_OnPlaySpaceFinished(GameObject
floorPlane)
    {
        InitAnchor();
    }
```

This is the handler assigned to the `OnPlaySpaceFinished` event raised by the `PlaySpaceManager` once it has finished scanning. Here, we are delegating the task to the `InitAnchor` method, which will be responsible for creating the anchor or for requesting it, depending on whether the instance is acting as a server or client.

Next, locate the `NetworkManager_OnNetworkReady(PWNetworkManager networkManager)` method and replace the `// TODO` comment with this code snippet:

```
    private void NetworkManager_OnNetworkReady(PWNetworkManager
networkManager)
    {
        if (networkManager.IsServer)
        {
            NetworkServer.RegisterHandler(MsgRequestAnchor,
OnRecievedRequestForAnchor);
        }
        else
        {
            networkManager.client.RegisterHandler(MsgAnchor,
OnRecievedAnchorPacket);
        }
        InitAnchor();
    }
```

This is the delegate that's called when the `NetworkManager` raises the `OnNetworkReady` event. If this instance is acting as the server, we register a handler to handle the `MsgRequestAnchor` network messages, that is, the message sent by the client when requesting the anchor, otherwise we register the `OnRecievedAnchorPacket` handler to handle network messages of the `MsgAnchor` type (the handler responsible for assembling the serialized and chunked `WorldAnchor`). Lastly, we call `InitAnchor`, which we will look at now. Locate this method and replace the `// TODO` comment with the following:

```
void InitAnchor()
{
    if (State != States.Undefined) { return; }

    if (networkManager.IsReady && networkManager.IsServer &&
PlaySpaceManager.Instance.Finished)
    {
        CreateAnchor();
    }
    else if (networkManager.IsReady && !networkManager.IsServer &&
PlaySpaceManager.Instance.Finished)
    {
        RequestAnchorFromServer();
    }
}
```

We first check the `State`; this state is updated during the process of exporting/importing and transmission. `Undefined` means the process has yet to begin and therefore it's okay to proceed. The first condition is server specific while the second is specific to the client; if the server is ready, we call `CreateAnchor` and if the client is ready, we call `RequestAnchorFromServer`.

Let's follow the server path first and then the client; locate the `CreateAnchor` method and replace the `// TODO` comment with the following code snippet:

```
void CreateAnchor()
{
    State = States.Exporting;

    if (PlaySpaceManager.Instance.Floor.GetComponent<WorldAnchor>
    ())
    {
    Destroy(PlaySpaceManager.Instance.Floor.GetComponent<WorldAnchor>
    ());
    }

    var worldAnchor =
    PlaySpaceManager.Instance.Floor.AddComponent<WorldAnchor>();
```

```
        StartCoroutine(ExportFloorAnchor(worldAnchor));
    }
```

We first update the `State`, and then destroy any preexisting anchor (alternatively, you can use this). Next, we add a `WorldAnchor` to the `PlaySpaceManager.Instance.Floor` and call a coroutine to export the newly created `WorldAnchor`. Let's continue with following the flow of the code; locate the `ExportFloorAnchor(WorldAnchor worldAnchor)` method and replace the `// TODO` comment with the following code:

```
    IEnumerator ExportFloorAnchor(WorldAnchor worldAnchor)
    {
        while (!worldAnchor.isLocated)
        {
            yield return new WaitForSeconds(0.5f);
        }

        WorldAnchorTransferBatch transferBatch = new
WorldAnchorTransferBatch();
        transferBatch.AddWorldAnchor(gameObject.name, worldAnchor);
        WorldAnchorTransferBatch.ExportAsync(transferBatch,
OnExportDataAvailable, OnExportComplete);
    }
```

In this first part, we iterate indefinitely until the `WorldAnchor` is located; this can be the case while the `WorldAnchor` is still initializing itself. Once it has been located, we begin the process of exporting, passing in our `OnExportDataAvailable` and `OnExportComplete` delegates to capture the data and be notified once the process is complete.

Most of this should look familiar to you from the last chapter. Within the `OnExportDataAvailable(byte[] data)` method, we simply copy the received data into our `worldAnchorBuffer` buffer and once complete, we check whether the process was successful and, if so, update the state. To finish off exporting, jump to the `OnExportComplete(SerializationCompletionReason completionReason)` method and replace it with the following code snippet:

```
    private void OnExportComplete(SerializationCompletionReason
completionReason)
    {
        if (completionReason == SerializationCompletionReason.Succeeded)
        {
            State = States.Anchored;
        }
        else
        {
            Debug.LogErrorFormat("Failed to export Anchor; {0}",
completionReason.ToString());
```

```
        }
    }
```

As mentioned in the preceding code snippet, if successful, we update the state to `Anchored`. This concludes exporting; staying with the flow of the server, let's now turn our attention to the transmission of the anchor to the client.

As shown in the preceding code snippet, the server listens out for the `MsgRequestAnchor` network message, which is sent by the client when wanting the anchor. Still within the `AnchorSyncManager` script, locate the handler assigned to the `OnRecievedRequestForAnchor(NetworkMessage netMsg)` message:

```
void OnRecievedRequestForAnchor(NetworkMessage netMsg)
{
    StartCoroutine(SendAnchorToClient(netMsg.conn));
}
```

Here, we are simply delegating the task to a coroutine, `SendAnchorToClient(NetworkConnection conn)`, the next method we will implement.

You may recall, from the last chapter, how large these anchors can be. Like with any programming problem, our choice of solution is vast, but here we have taken the simplest route in the hope that it will help you get more familiar with Unity's networking API.

Unity's HLAPI uses messages to encapsulate data sent between peers on the network; you may have noted the reference to messages mentioned earlier. All messages derive from the `MessageBase` class, and Unity provides a generic set of message types along with handling the heavy lifting of serializing and deserializing them. For this task, we will be creating a bespoke message, which we can use to transmit chunks of the exported `WorldAnchor` across the network to the client.

Locate the `AnchorDataMessage` class either with the Unity Editor or Visual Studio, and open it by double-clicking on the file. Once opened, you will be presented with an empty class, as shown:

```
using UnityEngine.Networking;

public class AnchorDataMessage : MessageBase
{
    public AnchorDataMessage()
    {

    }
}
```

This message is very rudimentary, but provides a nice opportunity to create one. As mentioned in the preceding code snippet, our message simply needs to transport across chunks of the `WorldAnchor` to the client. Along with this, we will add a packet number to allow us to reassemble it in order on the other end as well as a terminator flag to indicate the last packet. Let's add these variables now; add the following variables to the `AnchorDataMessage` class:

```
public int packetNumber;
public bool isEnd = false;
public byte[] data;
```

With our message class now complete, return to the `SendAnchorToClient(NetworkConnection conn)` method of the `AnchorSyncManager` class and replace the method with the following code snippet:

```
IEnumerator SendAnchorToClient(NetworkConnection conn)
{
    const int maxPayloadSize = 2048;

    AnchorDataMessage netMsg = null;
    int messageNumber = 0;
    int si = 0;
    int messagesSent = 0;

    var data = new List<byte>(Compress(worldAnchorBuffer.ToArray()));

    while (si < data.Count)
    {
        int count = Mathf.Min(maxPayloadSize, (data.Count - si));
        byte[] payloadData = data.GetRange(si, count).ToArray();

        netMsg = new AnchorDataMessage();
        netMsg.packetNumber = messageNumber;
        netMsg.data = payloadData;

        conn.SendByChannel(MsgAnchor, netMsg, 2);
        messagesSent += 1;
        messageNumber += 1;
        si += payloadData.Length;
        if (messagesSent % 20 == 0)
        {
            yield return new WaitForSeconds(0.3f);
        }
        else
        {
            yield return null;
        }
```

```
    }

    netMsg = new AnchorDataMessage();
    netMsg.packetNumber = messageNumber;
    netMsg.isEnd = true;
    netMsg.data = new byte[] { 0 };
    conn.SendByChannel(MsgAnchor, netMsg, 2);
}
```

There is a bit going on, so we will walk through the method by pulling out relevant snippets. First, we set up some variables, including a constant--maxPayloadSize--that determines the largest amount of data we will bundle into a message. Next, we compress the data and cast it into a List array for convenience when accessing the data:

```
var data = new List<byte>(Compress(worldAnchorBuffer.ToArray()));
```

We then iterate through the buffer array with a step size of maxPayloadSize or the remaining bytes, if smaller, as shown in the following snippet:

```
int count = Mathf.Min(maxPayloadSize, (data.Count - si));
```

Then, we pull out a chunk of data determined by the current index (si) and count determined in the preceding snippet, and use this to create a new message before sending it across to the client:

```
byte[] payloadData = data.GetRange(si, count).ToArray();

netMsg = new AnchorDataMessage();
netMsg.packetNumber = messageNumber;
netMsg.data = payloadData;

conn.SendByChannel(MsgAnchor, netMsg, 2);
```

Important to note is that we are sending the message using the channel created for this project (**Reliable Fragmented**), for reasons discussed earlier. Next, we update the variables, keeping a track of the packet number along with the index and then throttle the process, as illustrated in the given code snippet:

```
if (messagesSent % 20 == 0)
{
    yield return new WaitForSeconds(0.3f);
}
else
{
    yield return null;
}
```

We do this to ensure that we don't overwhelm either the sending queue of the server or receiving queue of the client; if we try to send messages too quickly, we are at risk of losing messages.

Once we have chunked and sent all the data, we send through one last message to inform the client that all the data has been received. Included in this message is the total number of messages sent (using the `messageNumber` variable); this allows the client to reason it whether it has received all the data or not:

```
netMsg = new AnchorDataMessage();
netMsg.packetNumber = messageNumber;
netMsg.isEnd = true;
netMsg.data = new byte[] { 0 };

conn.SendByChannel(MsgAnchor, netMsg, 2);
```

This completes the process of the server; let's now turn our attention to the client and how the client fetches and creates an anchor received from the server.

We left the client process at the point it was calling `RequestAnchorFromServer` from the `InitAnchor()` method. Let's now pick up from there and follow the client's flow. Locate the `RequestAnchorFromServer` method and replace it with the following code snippet:

```
void RequestAnchorFromServer()
{
    receivedAnchorDataMessages.Clear();
    State = States.RequestingAnchor;
    networkManager.client.Send(MsgRequestAnchor, new
IntegerMessage(MsgRequestAnchor));
}
```

Here, we simply clear the buffer, update the state, and send a message to let the server know that we are ready to receive the anchor.

As we saw earlier, the server, when it receives this message, will begin the process of sending through chunks of the anchor to the client via the `MsgAnchor` message type; a message to which we registered the `OnRecievedAnchorPacket` handler. Locate this method and replace it with the following code snippet:

```
void OnRecievedAnchorPacket(NetworkMessage netMsg)
{
    State = States.Downloading;

    var anchorPacket = netMsg.ReadMessage<AnchorDataMessage>();

    if (anchorPacket.isEnd)
    {
        numberOfPacketsExcepting = anchorPacket.packetNumber;
    }
    else
    {
        receivedAnchorDataMessages.Add(anchorPacket);
    }

    CheckForCompletedAnchor();
}
```

We first update the state to indicate that we are in the process of downloading; next, we cast the message to `AnchorDataMessage`. We then proceed to check whether it's the end packet or not, if it is, we set the `numberOfPacketsExcepting` variable, otherwise we append the `receivedAnchorDataMessages` collection with the data bundled with the message. We finally call the `CheckForCompletedAnchor` method. Locate and replace this method with the following:

```
void CheckForCompletedAnchor()
{
    if (numberOfPacketsExcepting == -1)
    {
        return;
    }

    if (receivedAnchorDataMessages.Count ==
    numberOfPacketsExcepting)
    {
        var data = receivedAnchorDataMessages
            .OrderBy((netMsg) => { return netMsg.packetNumber; })
            .Select((netMsg) => { return netMsg.data; })
            .SelectMany(netMsg => netMsg)
            .ToArray();

        data = Decompress(data);
```

```
        SetAnchor(data);
    }
}
```

In the preceding code, we are using the `numberOfPacketsExcepting` variable to determine whether we have received all packets. If it is not set (meaning that we are yet to receive the last message), we exit; otherwise we check this value with the number of packets received and if equal, we sort the received messages in order of `packetNumber` before creating a byte array for decompression. Finally, pass across to the `SetAnchor` method, which we will look at next. Locate `SetAnchor` and replace it with the following code snippet:

```
bool SetAnchor(byte[] data)
{
    State = States.Importing;

    worldAnchorBuffer.Clear();
    worldAnchorBuffer.AddRange(data);
    WorldAnchorTransferBatch.ImportAsync(data, OnImportComplete);

    return true;
}
```

We update the `State` before attempting to import the `WorldAnchor` via a `WorldAnchorTransferBatch`, passing the `OnImportComplete` delegate to be notified when the job is finished, which happens to be our final method. Let's finish this class off by replacing the `OnImportComplete` method with the following code snippet:

```
private void OnImportComplete(SerializationCompletionReason
completionReason, WorldAnchorTransferBatch deserializedTransferBatch)
    {
        if (completionReason ==
        SerializationCompletionReason.Succeeded)
        {
            if
            (PlaySpaceManager.Instance.Floor.GetComponent<WorldAnchor>
            ())
            {
Destroy(PlaySpaceManager.Instance.Floor.GetComponent<WorldAnchor>());
            }

deserializedTransferBatch.LockObject(PlaySpaceManager.Instance.Floor.name,
PlaySpaceManager.Instance.Floor);

            State = States.Anchored;
        }
        else
```

```
        {
            Debug.LogErrorFormat("Failed to import Anchor; {0}",
            completionReason.ToString());
        }
    }
```

If successful, we transfer the `WorldAnchor` to the client's `Floor` object before updating its state as `Anchored`. If we inspect the `State` property, we can see that changes to the state are broadcast via the `OnStateChanged` event. This is one of the events our `AppManager` class listens out for when coordinating the tasks of the game. Let's wrap this chapter up by completing the implementation of `AppManager`, the topic of the next section.

Coordinating

In this section, we will tie everything together by completing the `AppManager`; as implied by the title of this section, the `AppManager` is concerned with coordinating the activities of the game and ensuring that dependencies are fulfilled before starting a new task. In this project, the `AppManager` is also responsible for the spawning of the player entities, which is the main focus of this section. With that being said; let's jump into it and start putting things in place; open the script by either double-clicking on the file from within the Unity Editor (located in the /`PaintWarz/Scripts` folder within the **Project** panel) or from within Visual Studio.

Within the `AppManager` class, we have assigned all relevant prefabs accessible via the public variables:

```
public GameObject playerSPrefab;
public GameObject playerCPrefab;
```

Within the AppManager Start method, we register to events associated with the `PlaySpaceManager` finishing scanning, `AnchorSyncManager` state updating, and when the user performs a tap via the `InputManager`; the assigned handlers are probably a good place to start our conversation. Let's first preview the `Start` method, where these handlers are assigned:

```
void Start () {
    ...
    networkManager.OnNetworkReady += NetworkManager_OnNetworkReady;
    networkManager.OnConnectionOpen += NetworkManager_OnConnectionOpen;

    PlaySpaceManager.Instance.OnPlaySpaceFinished +=
  PlaySpaceManager_OnPlaySpaceFinished;
```

```
        AnchorSyncManager.Instance.OnStateChanged +=
    AnchorSyncManager_OnStateChanged;

        InputManager.Instance.OnTap += InputManager_OnTap;
    }
```

Let's work our way through each handler, starting with
the `NetworkManager_OnNetworkReady` method. To recap, the `NetworkManager` raises the
`OnNetworkReady` event when either the server is ready to receive connections or the client
has connected to the server, while the `OnConnectionOpen` event is raised when a
connection is made between the server and the client. Find and replace the
`NetworkManager_OnNetworkReady` method with the following code snippet:

```
    void NetworkManager_OnNetworkReady(PWNetworkManager networkManager)
    {
        if (networkManager.IsServer)
        {
            NetworkServer.RegisterHandler(MsgTeamSpawnRequest,
    OnRecievedRequestToSpawnTeam);
        }
    }
```

If this instance is acting as a server, we register the `OnRecievedRequestToSpawnTeam`
handler for the `MsgTeamSpawnRequest` message type. Before continuing, let's take a quick
detour to discuss the concept of authority.

You may recall a previous discussion about using a single point of control (or authority) to
simplify the synchronization and control of networked objects. Unity supports this through
the concept of authority, whereby only the instance with the authority of a specific object
can update its properties and, once updated, will send through the updated state to its
ghost self within the other instances. An error will be raised if the instance doesn't have
authority to modify its properties.

The following figure illustrates this concept, where a single instance *owns* the object and a representation, or ghost, exists on the other instances:

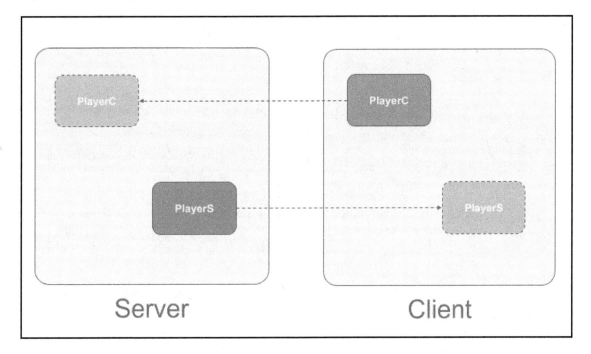

You may be wondering what exactly a networked object is, is it just a plain old GameObject? Almost, it's a GameObject with a NetworkIdentity, which is responsible for creating a unique identifier for an object across the network; for example, in the preceding image, the instance of PlayerS on the server would have the same identify as the PlayerS object on the client, along with details of which instance owns it and so on. For more details, refer to the official documentation at https://docs.unity3d.com/Manual/class-NetworkIdentity.html.

Before jumping back into the code, it's worth noting that only the server can spawn networked objects; it is for this reason that the client sends the MsgTeamSpawnRequest message.

Let's now return our focus to the task at hand and resume where we left off, with implementing the `OnRecievedRequestToSpawnTeam` method; replace this method with the following code snippet:

```
void OnRecievedRequestToSpawnTeam(NetworkMessage netMsg)
{
    var teamSpawnRequestMessage =
netMsg.ReadMessage<TeamSpawnRequestMessage>();

    CreateTeam(netMsg.conn, teamSpawnRequestMessage.position);
}
```

When the server receives this message, it first deserializes the message to an instance of `TeamSpawnRequestMessage` before calling `CreateTeam`, and passing the connection and position variable bundled with the received message.

 Just to clarify, this method is only called on the server, so we currently reside within the instance of the server, satisfying a request by the client.

Find and replace the `CreateTeam` method with the following code snippet:

```
public void CreateTeam(NetworkConnection connection, Vector3
teamCenterPosition)
{
    const int teamSize = 3;
    string team = connection == null ? "TeamS" : "TeamC";

    for (int i = 0; i < teamSize; i++)
    {
        GameObject player = null;
        if (connection == null)
        {
            player = Instantiate(playerSPrefab);
        }
        else
        {
            player = Instantiate(playerCPrefab);
        }
        player.GetComponent<Player>().Init(team,
string.Format("Player_{0}_{1}", team, i));
        var playerOffset = UnityEngine.Random.insideUnitSphere * 0.5f;
        playerOffset.y = 0;
        player.transform.position = teamCenterPosition + playerOffset;
        if (connection == null)
        {
```

```
                    player.GetComponent<NetworkIdentity>().localPlayerAuthority
= false;
                    NetworkServer.Spawn(player);
        }
        else
        {
                    player.GetComponent<NetworkIdentity>().localPlayerAuthority
= true;
                    NetworkServer.SpawnWithClientAuthority(player, connection);
        }
    }
  }
```

We first determine whether this is for a client or server; from this, we instantiate the appropriate prefab. Next, we iterate through a loop of teamSize to create the desired size squad. For each player, we apply a little jitter to the position to avoid having each one spawned on top of another. The final, and most critical part, is spawning the instance across the network. If the connection is null, then this call was for the server; therefore, the authority resides with the server, in which case we use NetworkServer.Spawn(player) to spawn the player. Otherwise, we call NetworkServer.SpawnWithClientAuthority(player, connection), passing in the connection that hands over the authority to the client, that is, the client is responsible for updating this object.

At this stage, we have seen how the server handles a request to create a team; what are remaining are the actions required to create that message. Creating a team is triggered by the user performing a tap gesture when gazing at the floor; jump back to the InputManager_OnTap(GameObject target, Vector3 hitPosition, Vector3 hitNormal) method and replace it with the following code snippet:

```
    private void InputManager_OnTap(GameObject target, Vector3 hitPosition,
  Vector3 hitNormal)
    {
        if (!teamCreated &&
  PlaySpaceManager.Instance.IsCloseToFloor(hitPosition))
        {
            InitTeam(hitPosition);
        }
        else
        {
            if (!string.IsNullOrEmpty(Team) &&
  PlaySpaceManager.Instance.IsCloseToFloor(hitPosition))
            {
                tapIndicator.Show(hitPosition, hitNormal);
                TeamsManager.Instance.SetTeamsTarget(Team, hitPosition);
            }
        }
```

```
        }
    }
```

When a tap gesture is detected, we first determine whether a team has been created; if not, we call `InitTeam`, which handles the process of creating the team. If a team has already been created, we update the team's position via the `TeamsManager.Instance.SetTeamsTarget` method, passing in the team's name and target position.

We are almost finished; at this point, we have an application that scans the environment, discovers and connects with its pairs, shares the coordinate system, and instantiates the team of the player's squad. The last task is to parent each of the players to the anchored floor. This means that any adjustment to the floor will have a consequence for the children, a desirable side effect. We will perform this with the space that has been scanned; find and replace the `PlaySpaceManager_OnPlaySpaceFinished(GameObject floorPlane)` method with the following code snippet:

```
    private void PlaySpaceManager_OnPlaySpaceFinished(GameObject
floorPlane)
    {
        var players = FindObjectsOfType<Player>();
        foreach (var player in players)
        {
            player.transform.parent = floorPlane == null ? null :
floorPlane.transform;
        }
    }
```

Here, we simply find all objects of the `Player` type, and set their parent to the anchored floor, as discussed earlier. That's it! This concludes this project, and now is a perfect opportunity to test it. Grab a friend nearby with the device and build and deploy to test:

Summary

In this chapter, our goal was to explore collaboration on the HoloLens further; here, our focus was to introduce some of the networking components made available by Unity. We started our journey by discussing the core components that facilitate and support a multiplayer networked game. We then walked through the example project, illustrating these concepts through code, from discovering peers on a local network to creating a shared coordinate space and spawning new networked objects.

As with many of the chapters; we have only just dipped our toes into the world of networked entertainment, but hopefully, we provided enough incentive to motivate you to explore the topic further and continue with building new experiences that encourage us, as people, to engage with each other.

In the next and the final chapter, we will wrap up this book by looking at how to build and deploy HoloLens applications.

9

Deploying Your Apps to the HoloLens Device and Emulator

In this final (and brief) chapter, we will wrap up by discussing the details of deploying your application to the emulator and device. Whether you are developing using Unity or DirectX, the path is similar once the project has been exported to a Visual Studio project from Unity and most of our discussion will be focused on this. But before this, we will cover the Holographic Emulation and Holographic Remoting built into Unity. In this chapter, you will learn about the following:

- Using the Holographic Emulation and Holographic Remoting built into Unity
- Deploying and making use of the HoloLens emulator
- Deploying to the device
- Windows Device Portal

Rapid feedback with Unity's Tooling

I'm certain I'm not unique in my coding style--where at certain points, I find myself quickly iterating through the rapid development > deployment > testing > development > deployment > ... cycle. It is for this reason I was delighted when Unity added Holographic Simulation right inside Unity. I was similarly delighted, with remoting, which allowed even higher testing fidelity with less friction that would otherwise be required. In the following sections, we will discuss each in turn and then move on to deploying to the HoloLens emulator and the HoloLens device itself from Visual Studio.

Holographic Simulation

Agile development is, in essence, about reducing the feedback time, and the Holographic Simulation does just that. This allows rapid development and encourages experimentation, something a new platform, such as the HoloLens, needs. To get it up and running is simply a matter of turning it on and clicking on the **Play** button. This is done by first opening the **Holographic** panel by clicking on the menu via **Window | Holographic Emulation**, as shown:

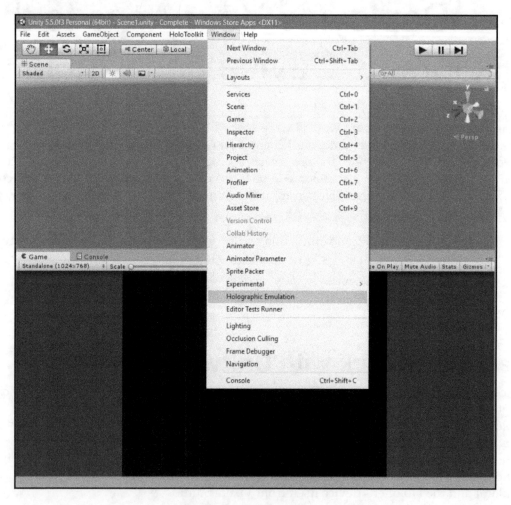

Steps to reach Holographic Emulation

You will now be presented with the **Holographic** panel; from here, it's simply a question of selecting **Simulate in Editor** for **Emulation Mode**, as shown in the following screenshot:

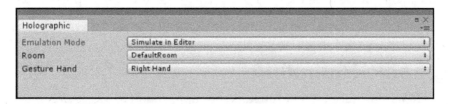

When **Emulation Mode** is set to **Simulate in Editor**, clicking **Play** will start the emulator built into the Unity Editor. The **Holographic** panel also exposes the **Room** and **Gesture Hand** options. **Room** allows you to select one of the supplied virtual rooms (located in the `Plugins/Rooms/` folder where the emulator resides) while **Gesture Hand** allows us to select between left-and-right handed gestures.

To take advantage of the simulator, you will need to have access to a game controller, such as an Xbox controller. You can still run the simulator but won't be able to move around.

The following table lists the controls available when the simulator is running; these were taken from the official Unity website:

Control	Function
Left stick	Up and down to move the virtual human player backward and forward; left and right to move left and right.
Right stick	Up and down to pitch the virtual human player's head up and down; left and right to turn the virtual human player left and right.
D-pad	Move the virtual human player up and down or roll head left and right.
Left and right trigger buttons; **A** button	Perform a tap gesture with a virtual hand.
Y button	Reset the pitch and roll of the virtual human player's head.

The simulator is great for rapid development and provides the opportunity to eliminate any obvious bugs that would otherwise be deployed to the device, adding to the friction. But it does lack the fidelity of wearing the device, which brings us nicely to the next tool released by Unity--Holographic Remoting.

Holographic Remoting

Holographic Remoting allows you to stream your application to the device. It behaves as if it was running on the device while skipping the time-consuming build and deployment processes.

To use remoting, you must first download and install the companion HoloLens application *Holographic Remoting Player*, available at `https://www.microsoft.com/en-us/store/p/holographic-remoting-player/9nblggh4sv40`. Once installed, launch the application, where you will be presented with a screen similar to the following screenshot:

With the HoloLens running the remoting player, return to the Unity Editor and select **Remote to Device** from the **Emulation Mode** drop-down on the **Holographic** panel. Enter the IP address shown in HoloLens into the **Remote Machine** field and click **Connect**, as shown in the following screenshot:

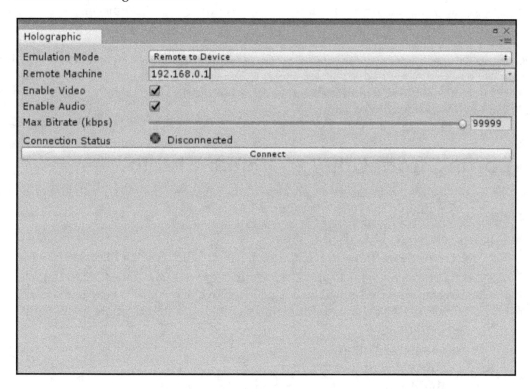

If successful, **Connection Status** will turn green and you are ready to go. With the device now connected, every time you click on **Play** within the Unity Editor, your application will be streamed to the HoloLens device. You can pause, inspect, and debug as you would normally do while running your game in the Editor.

Some known limitations taken from the official Unity website include:

- Speech (`PhraseRecognizer`) is not supported via remote to the device; instead, it intercepts speech from the host machine running the Unity Editor
- While remote to device is running, all the audio on the host machine redirects to the device, including the audio from outside your application

Their simplicity, convenience, and speed make these tools a necessity when developing for the HoloLens. Next, we will discuss deploying to the emulator and device from Visual Studio; but before this, we will quickly summarize building a HoloLens from Unity, which will result in a Visual Studio solution.

Exporting from Unity to Visual Studio

Because we have already discussed these steps in the previous chapters, I will just provide a quick summary outlining the steps required:

1. In Unity, select **File | Build Settings**.
2. Select **Windows Store** on the **Platform** list and click **Switch Platform**.
3. Set **SDK** to **Universal 10** and **UWP Build Type** to **D3D** - Direct3D (D3D) selected means the application will be hosted in a Direct3D window as opposed to the alternative, XAML, of a XAML window, making it slightly more lightweight and more performant.
4. Check **Unity C# Projects**.
5. Click **Add Open Scenes** to add the scene.
6. Click **Build** and select the destination folder.
7. When Unity is done, a file explorer window will appear where the project directory resides.

8. Open the folder and double-click on the `.sln` file to open the solution in Visual Studio.

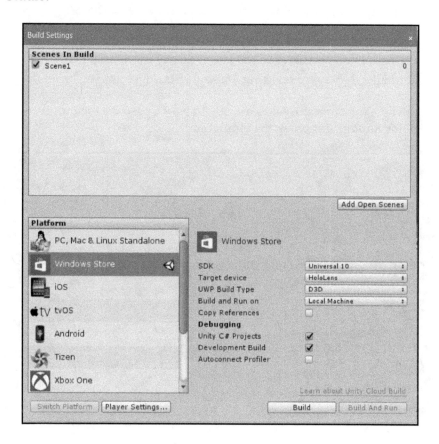

HoloLens emulator

The HoloLens emulator is a great intermediary for the actual device, allowing developers without a device to get involved and providing a more convenient workflow while developing.

The emulator exposes the inputs from what would normally come from the sensors to the peripherals and the virtual controls on the desktop, allowing the user to simulate gestures using a keyboard and mouse or an Xbox controller.

Deploying to the emulator

Perform the following steps to deploy your application to the emulator:

1. From within Visual Studio, set the **Platform** to **x86**.
2. Select **HoloLens Emulator** as the target device for debugging.
3. And finally, deploy by selecting **Debug** | **Start Debugging** from the menu.

When first launched, the emulator takes a while to boot up, so it's recommended to leave it running while developing and using the emulator:

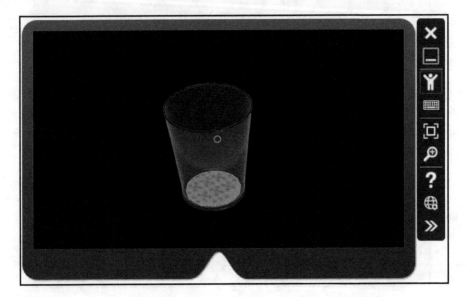

I have found that the easiest way to navigate around in the emulator is using an Xbox controller but it's possible to use the keyboard and mouse, as mentioned in the following section:

Moving around the environment (left, right, forward, and backward)	Use *A*, *D*, *W*, and *S* on your keyboard or the left stick of the Xbox controller
Looking around (look left, right, up, and down)	Use the arrow keys, or left-click and drag the mouse pointer or the right stick of the Xbox controller
Air tap gesture	Right-click on the mouse, *Enter* on the keyboard, or the **A** button on the Xbox controller

Bloom gesture	The Windows *Home* button, or *F2* on the keyboard, or the **B** button on the Xbox controller
Hand movement	Either hold the *Alt* key on the keyboard and right-click and drag your mouse cursor or hold the right trigger and the **B** button on the Xbox controller and move the right stick

The emulator also exposes a window to track input to the HoloLens and allow you to change the virtual room. You can access (and hide) this panel by clicking on the **Tools** button on the docked toolbar, as shown in the following screenshot:

The **Simulation** panel in the following screenshot shows the current state of the simulated sensors used to drive the HoloLens emulator. Hovering over any value in the **Simulation** tab will provide a tooltip describing how to control that value:

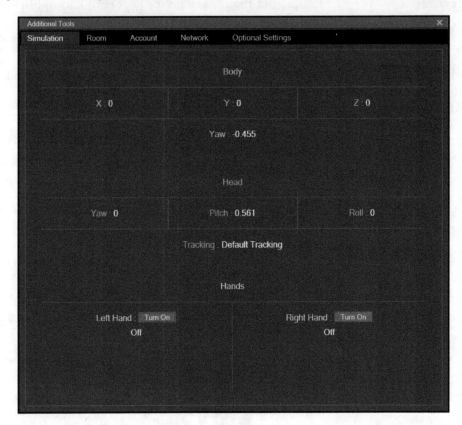

Simulation panel

The next panel, **Room**, allows you to load a different (virtual) environment to test your application in. Access it by simply clicking on the **Room** tab and then clicking on the **Load Room** button. This will open a file dialog residing in the default directory where the rooms are stored. Selecting a different room will result in that room being loaded into the emulator.

Microsoft has provided us with a way to easily create new environments, something we will briefly cover in the **Windows Device Portal** section later on. You can access the Device Portal using a web browser. The IP address that you use is made available in the **Network** panel under **Emulator Adaptor #1 | Network addresses**, as shown in the following screenshot or, alternatively, by clicking on the global button on the toolbar docked to the emulator:

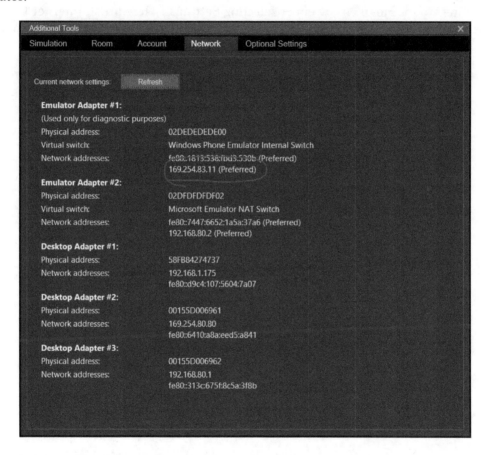

And with that, we have finished the section specifically about the emulator. Next, we move on to deploying to the device. For information about the emulator, check out the official website `https://developer.microsoft.com/en-us/windows/mixed-reality/using_the_hololens_emulator`.

Deploying to the device

The process of deploying to the device is almost identical to deploying to the emulator. Back in Visual Studio, update the deployment destination from **HoloLens Emulator** to **Remote device**. When selected for, the first time, you will be presented with a dialog requesting the IP address of the remote device--the HoloLens in our case. You can obtain the IP address of the HoloLens by either selecting **Settings** | **Network & Internet** | **Advanced Options** from the main menu on the device or asking Cortana "Hey Cortana, what's my IP address?" (Also on the device).

 You can also deploy to the HoloLens over USB; details are omitted here due to the convenience offered by Wi-Fi.

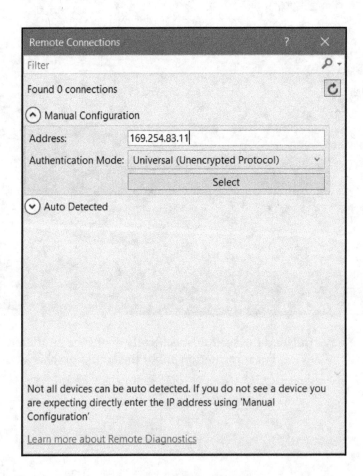

Finally, close the **Remote Connections dialog** by clicking on the **Select** button and deploy to the device by selecting **Debug | Start Without debugging** from the menu.

If this is the first time you are deploying to your device from the computer, you will need to pair them. If this is the case, then the HoloLens will display a PIN along with Visual Studio, presenting a prompt for you to enter it into. You can also pair manually by having the HoloLens generate a PIN, by launching **Settings | Update | For Developers | Pair**. A PIN will be displayed on the HoloLens, with Visual Studio presenting a prompt for the PIN to be entered into. Once entered, tap on the **Done** button on your HoloLens to dismiss the dialog. Once paired, this step is ignored for subsequent deployments.

This concludes deployment. We will now briefly discuss the Windows Device Portal before wrapping up.

Windows Device Portal

The Windows Device Portal for HoloLens lets you configure and manage your device remotely over Wi-Fi or USB. The device portal is a web server that running on the HoloLens that can be accessed from the web browser. In the previous section, we discussed deploying to the emulator and device. In each case, we obtained the IP address. We can use this to access the device Portal; simply enter the IP address obtained from the HoloLens device or emulator into a web browser. No authentication is required when logging in to the emulator unlike logging in to the device. The details of these steps can be found on the official website, `https://developer.microsoft.com/en-us/windows/mixed-reality/using_the_windows_device_portal#connecting_over_wi-fi`.

The Device Portal exposes a lot of functionality to configure and manage the device. Details of this can be found on the page mentioned in the previous paragraph, but the intention here is just to highlight its existence and usefulness when developing. At the start of this chapter, I emphasized the importance of the feedback loop when developing. The quicker you can test and iterate, the quicker you can explore, experiment, and of course, squash bugs. One useful feature offered is the ability to create your own rooms. You may do this to:

- Better understanding the space you are designing and developing for
- Align more effectively with the environment your application will be used in

Rooms (`/environments`) can be easily created by having the device on and connected to the Device Portal, selecting **3D View** on the left pane within the Device Portal. This will present a page that allows you to see what the HoloLens sees. Clicking on **Update** will update the spatial mapping the HoloLens has made of the current space--this can be downloaded as a `Wavefront.obj` file and imported into most 3D modelling packages, such as Blender 3D. This view is shown in the following screenshot:

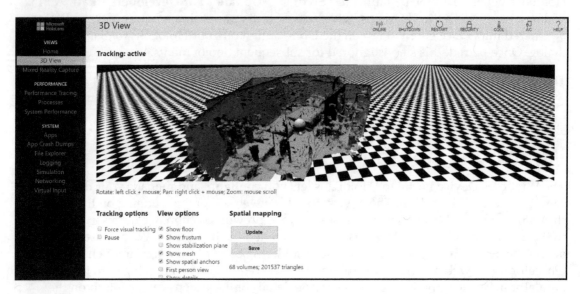

Your next step is to export it out to a format that emulator can load. This has the extension `.xef` and is a format borrowed from Microsoft's Kinect. To export your room, navigate to the Simulation page by clicking on the **Simulation** link on the left pane. The next step is just a matter of giving your room a name in the **Room Name** field and clicking on the **Capture** button, as shown in the following screenshot:

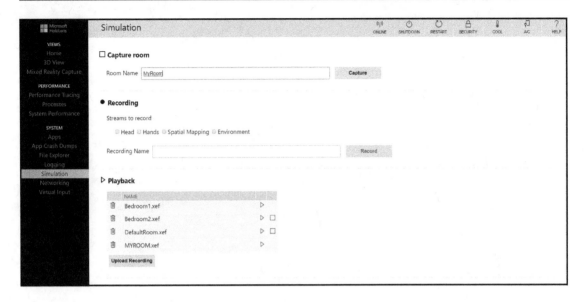

Once you do this, a `.xef` file will be downloaded to your disk. To make use of this file, return to the emulator, open the **Tools** window, select the **Room** panel, and load in the downloaded `.xef` file.

The Device Portal offers much more than I have discussed here but I really just wanted to highlight its availability and ability to create and download your own 3D models for testing within the Editor as I have found this invaluable during development. You can learn more about the Device Portal by visiting the official web page `https://developer.microsoft.com/en-us/windows/mixed-reality/using_the_windows_device_portal#connecting_over_wi-fi`.

Summary

And that's a wrap. I hope you have enjoyed our journey during which we have developed a range of mixed reality applications, ranging from the Terminator-style vision to collaborative design tools. As mentioned throughout this book, the technology is very much in its infancy, but I hope the examples and discussions presented here have got you excited about the near future, one where our computers are not boxes that sit on our desks but are augmentations of ourselves. With that in mind, if you could give someone superpowers, what would they be?

Index

www.ingramcontent.com/pod-product-compliance
Lightning Source LLC
LaVergne TN
LVHW081512050326
832903LV00025B/1456